ANALYTICAL BIOCHEMISTRY

ANALYTICAL BIOCHEMISTRY

A Practical Handbook for Undergraduates

MARK C. HALL

PURDUE UNIVERSITY PRESS
WEST LAFAYETTE, INDIANA

Cataloging-in-Publication Data on file at the Library Congress.

978-1-62671-386-4 (paperback)

978-1-62671-389-5 (epdf)

Cover images: DNA double helix with sparkling nodes in a dynamic blue environment. Genomic sequencing, molecular biology, and genetic data analysis: TanyaJoy/iStock via Getty Images Plus; digital landscape with dots and lines. Cyberspace grid: Ihor Tsyvinskyi/iStock via Getty Images Plus; amino acid chemical molecules and chain structures: seamartini/iStock via Getty Images Plus; atomic model of a protein structure: Generated by the author using AlphaFold2 by Google LLC, and Molecular Operating Environment by Chemical Computing Group. For more information about AlphaFold Servers, please visit https://www.nature.com/articles/s41586-024-07487-w or https://alphafoldserver.com/welcome. For more information about Chemical Computing Group, please visit https://www.chemcomp.com/en/Products.htm.

Any figures available in color appear that way in the e-book format.

CONTENTS

Accompanying instruction guides are available online at https://docs.lib.purdue.edu/analyticalbiochemistry/

PREFACE

PURPOSE

Protein purification and characterization skills are important in numerous biochemistry-related career areas, including basic research, and in pharmaceutical and biotech industries, such as medical, agricultural, and industrial biotech. This book is intended as a comprehensive guide for undergraduate students taking an analytical biochemistry lab course focused on modern practical and critical thinking skills related to protein purification and characterization. The practical and intellectual skills taught and practiced in this course are intended to help prepare students for a broad array of careers, including scientist positions in the industries listed above as well as advanced graduate education in biochemistry and the molecular life sciences.

The context for the course is the early target validation and lead identification stages of the drug discovery process. As such, the course project relates directly to common objectives encountered in industry settings. In addition to teaching methods for protein purification and characterization, the course provides practice in applying the scientific method, including hypothesis development and testing, as well as writing scientific reports. The course therefore seeks to train students in the process of scientific inquiry and discovery. Foundational biochemistry principles, such as protein structure and function and enzyme catalysis, are emphasized and highlighted throughout the course to provide clear connections between theoretical concepts and their real-world relevance and applications.

Importantly, the format for this course is different than a traditional lab course in which activities from one week to the next are often unconnected. This course is designed to simulate a research project that lasts the entire semester. There is a single overarching objective, and all activities are related in some way to accomplishing it. The expectation is that this format better introduces students to the process of scientific research, helps convey purpose for various protein analytical methods, and provides a better opportunity for practicing the scientific method and scientific communication as they would be used in real research.

While this book is specifically designed as a guide for students in the BCHM 32200, Analytical Biochemistry, course at Purdue University, the hope is that it is flexible enough to be used more generally to support biochemistry lab courses with similar goals at other undergraduate educational institutions. While some specialized lab equipment is required, most of the lab experiments and activities can be accomplished on a minimal budget with common instructional lab instrumentation and supplies and relatively inexpensive and standard reagents. In the accompanying instruction guide, suggestions are made for when alternative procedures, reagents, supplies, or equipment could be substituted for the specific items mentioned in the protocols.

TARGET AUDIENCE

This book is specifically written as a practical guide for undergraduate students enrolled in BCHM 32200, Analytical Biochemistry, at Purdue University. The book contains the weekly instructions and covers related theoretical and foundational principles for the lab activities in this course. However, this book can also serve as a blueprint at other institutions for similar undergraduate biochemistry lab courses that share the same focus, objectives, and learning outcomes.

Students in BCHM 32200 are typically juniors, and this course is the final biochemistry lab course required in the biochemistry BS plan of study at Purdue. Thus, students are expected to have already mastered general chemistry and organic chemistry principles and have a working knowledge of some basic biochemistry lab skills, including (but not limited to) use of micropipettors and the ability to make buffered solutions with specific concentrations of ingredients at a specific pH. Basic math skills are important as well, as is a working knowledge of Microsoft Excel spreadsheets for data processing and graphing. While fundamentals of protein structure and function are covered and reviewed in this course, it is expected that students have already acquired this knowledge at an introductory college level.

LEARNING OBJECTIVES

The course is designed to teach the practice of science in addition to teaching common contemporary analytical biochemistry methods. The following learning objectives are targeted:

1. Give students practical experience with the process of isolating and characterizing biochemical properties of proteins and enzymes.
2. Expose students to classical and modern methods employed in protein characterization, including computational tools used to augment wet lab experiments. Methods include site-directed mutagenesis, recombinant protein expression, affinity and size exclusion chromatography, electrophoresis, centrifugation, immunoblotting, spectrophotometry, enzyme activity assays, mass spectrometry, bioinformatics, and protein structural modeling. Concepts of molecular biology, such as creating recombinant DNA plasmids, are also covered but are not part of the lab activities.
3. Introduce students to the early stages of the drug discovery process, including target validation and lead compound identification.
4. Provide students with opportunities to develop hypotheses based on existing knowledge and to design appropriately controlled experiments to specifically test those hypotheses.
5. Give students practice effectively communicating scientific research results in writing.

LEARNING OUTCOMES

Students are expected to be able to do the following after completing this course. The methods used for assessment are listed after each outcome.

1. Students will be able to apply the scientific method to a specific research problem. This will include the design of hypotheses based on information found in primary literature, design of experiments to test hypotheses, and analysis of experimental data to determine if hypotheses are supported.
   Methods of assessment: Weekly lab data submissions, weekly lab question sets, final report.
2. Students will be proficient in the application of modern biochemical and molecular techniques for the purification and characterization of proteins.

Methods of assessment: Weekly lab instruction quizzes, weekly lab data submissions, weekly lab question sets, and midterm and end-of-semester exams.

3. Students will be able to explain the theoretical principles behind modern research methods for protein purification and characterization.
   Methods of assessment: Weekly lecture quizzes and midterm and end-of-semester exams.
4. Students will be able to explain the contributions of the course methods to society, especially to the fields of drug discovery and agricultural biotechnology.
   Methods of assessment: Weekly lecture quizzes and final report.
5. Students will be able to effectively document and communicate research results and their meaning in writing.
   Methods of assessment: Final report, including submissions of individual section drafts during the semester.

FORMAT OF THE COURSE

Many undergraduate lab courses focus on teaching a collection of methods with little to no connection between them and/or no underlying theme and objective. This course, in contrast, is structured to simulate a semester-long research project where a variety of methods for protein analysis are employed to complete the research objective. The project features the process of drug target validation, a critical early step in the drug development pipeline.[1] In this practical context, the importance and utility of each method should be readily apparent.

The course is structured with a one-hour lab prep period and a three-hour lab period, both on the same day of the week. In many weeks, the lab prep period is used to initiate experiments or procedures that are then completed in the full lab period. Lectures are also an important part of the course, providing theoretical background on the analytical methods used or covering related biochemistry principles. The BCHM 32200 course does not have a dedicated lecture period. Therefore, lectures are recorded and viewed by students asynchronously via Purdue's course learning management system. In general, the lecture topics cover methods or content related to a given week's lab activities. The asynchronous lecture format requires students to be self-motivated and diligent. The use of weekly quizzes on the lectures prior to the related lab period is intended to ensure that students fully prepare for each week. Likewise, weekly prelab quizzes on the lab activities are intended to ensure that students have read the lab instructions and understand the lab activities and their goals ahead of time. This hopefully helps the labs run more efficiently and reduces the failure rate of experiments.

The expectation is that students work in teams of two or three depending on class size and availability of equipment and supplies. Other group sizes are certainly possible, but the practical learning may be diluted as the group size increases because students are spending more time watching their teammates do the experiments instead of doing them with their own hands. Larger group sizes can also make it too easy to rely on others to do the troubleshooting, planning, and interpretation needed to successfully complete experiments. It is important that each student get the full practical educational experience.

The overall project will be described in Week 1 and will culminate with a final written report describing the collective experiments and results in the format of a typical peer-reviewed biochemistry research journal. As noted above, in this semester-long continuous research project the emphasis is as much on the process of science and critical thinking as it is on learning specific methods. The methods continue to be important as well but are learned in the context of helping answer an overarching question and achieving a collective goal.

In addition to the weekly quizzes aimed at student preparation for the labs, the following assessments exist to monitor learning outcomes and assign grades:

- A data score is assigned each week based on results and analyses.
- A set of thought questions must be answered after each lab and submitted by the following week.
- A midterm exam and end-of-semester exam (noncumulative) primarily assess material from lectures, background knowledge of the project and methods, and the ability to perform data analyses.
- The single manuscript-style report documenting the entire semester's efforts takes the place of a final exam. This report requires students to communicate the big picture significance of the topic, the rationale and objectives for the project, and the implications of their results within the context of an existing research area, in addition to describing their methods and results. Drafts of the various report sections are submitted at regular intervals during the second half of the course to prevent students from procrastinating and leaving an unmanageable workload at the end of the course. These drafts also allow feedback on general format and content to help students get started on the right track and improve their scientific communication skills.

At Purdue, the BCHM 32200 course is taught using the university's course learning management system, currently Brightspace. This is useful for making lecture videos available, automatically grading quizzes, collecting weekly question submissions, tracking grades, etc. As with most lab courses, this course should have at least one teaching assistant who can help set up labs and assist students during the lab activities, in addition to the primary instructor.

The following table contains the suggested weekly schedule for the course, with both lecture topics and lab activities listed:

TABLE P.1. *Schedule*

WEEK #	LECTURE TOPIC	LAB EXERCISES
1	Course overview and bioinformatics	Using bioinformatics tools (homology searching, sequence alignments, etc.)
2	Protein structure and function	Site-directed mutagenesis
3	Recombinant protein expression	Recombinant protein expression
4	Chromatography and protein purification	Protein affinity purification
5	Protein electrophoresis	Protein analysis by SDS-PAGE and measuring protein concentration
6	Using antibodies to study proteins	Quantitative immunoblotting
7	Spectroscopy and mass spectrometry	**Exam 1** and protein mass spectrometry
8	Protein hydrodynamic properties	Determining protein oligomeric state
9	Enzyme catalysis and regulation	Enzyme assays
10	Enzyme kinetics	Steady state enzyme kinetics
11	Molecular recognition	Substrate specificity
12	Enzyme inhibition	Enzyme inhibitor screening
13	Protein structure determination	Ligand docking and optimization
14	No lecture for second exam week	**Exam 2**
15	If there is an extra week it can be used for final report work	

INTRODUCTION: PROJECT OVERVIEW

DRUG DISCOVERY

The drug discovery pipeline is employed by pharmaceutical and agricultural biotech companies to generate new therapeutic compounds to treat diseases and pesticide compounds to protect crops from pests and diseases. Although these compounds are called pesticides in the agricultural world rather than drugs, their development process is fundamentally the same and follows the basic steps illustrated in Figure I.1. The term "drug" will be used generically in this chapter to refer to compounds used either as pharmaceuticals for treating human disease or as pesticides for treating plant disease.

Drug development typically begins with the identification and validation of a target molecule (often a protein and enzyme). In the case of a human disease, a suitable target for a drug is a protein that is required for the disease state. For an infectious disease the target would typically be an essential virulence factor in the pathogen. For example, the enzyme D,D-transpeptidase is essential for formation of the protective bacterial cell wall and is the target of the antibiotic drug penicillin (Figure I.2A).[1] Another example of a target is the enzyme HMG-CoA reductase, which catalyzes the rate-limiting step in the biosynthesis of cholesterol. Its inhibition by statin drugs such as lovastatin (Figure I.2B) is a very effective therapy for lowering cholesterol in patients suffering from hypercholesterolemia.[2] Both penicillins and statins were originally discovered as natural products produced by microorganisms but have been synthetically improved and diversified over the years using a variety of research tools that you will learn about in this course.

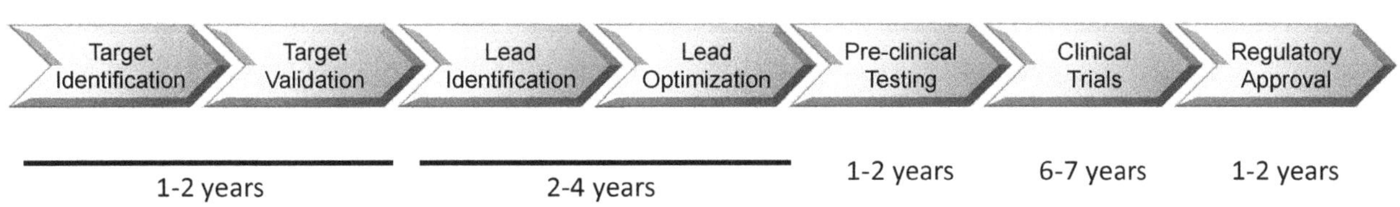

FIGURE I.1. The drug discovery pipeline begins with identification of potential targets linked to the disease state, followed by validation of target drugability. Subsequently, high throughput screens are established and conducted to identify promising drug leads. Iterative rounds of optimization through structural biology and medicinal chemistry generate a small number of drug candidates. The preclinical testing stage assesses candidate efficacy and safety in animal models. The most promising compounds undergo a rigorous three-phase clinical trial before final regulatory approval for commercialization and medical use. While the overall process is similar for chemical pesticides used in agriculture, the regulatory process is not as stringent.

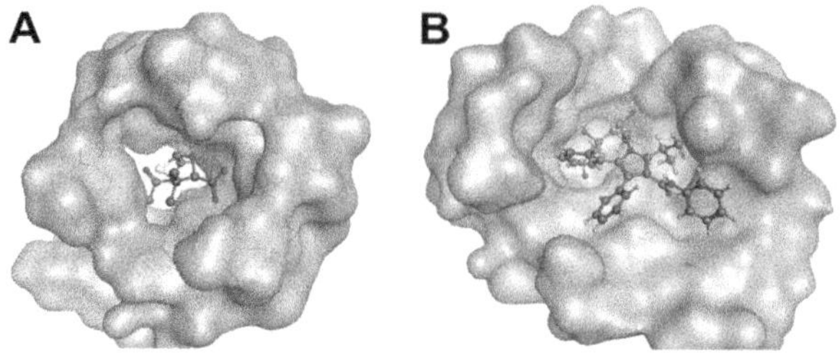

FIGURE I.2. (A) The active site of bacterial transpeptidase bound to a derivative of the antibiotic penicillin. (B) Similar view of the active site of human HMG-CoA reductase bound to the statin drug lovastatin. Both images are derived from the experimentally determined cocrystal structures of the enzyme-drug complexes, highlighting the intricate atomic interactions that define enzyme specificity, an important focus topic in this course. Images derived from the RCSB Protein Data Bank (PDB): PDB ID 5XCW (panel A) and 1HWK (panel B). Images generated with Molecular Operating Environment.

If not already validated, the target validation process would involve demonstrating that disrupting normal function of the target has a beneficial effect on the disease (human or crop). There are other practical considerations in selecting targets for drug or pesticide development. Not all proteins are easily "druggable," meaning molecules can be created that tightly and specifically bind to the target and can alter its function. Druggable proteins typically have well-defined ligand binding pockets where a drug molecule could bind, such as the active site of enzymes (see Figure I.2). It is usually helpful if the atomic structure of the target is already known, as this allows visualization of candidate drug molecules interacting with the target and facilitates optimization of drug structures using computational tools to maximize target binding affinity. Finally, if a target has many closely related proteins (in either the target species or other species), then serious unintended side effects may occur through unintended drug binding to the related proteins. These considerations are all part of the target validation process.

FUNGAL PATHOGENS IN AGRICULTURE

Plants, just like humans, can contract a variety of infectious diseases. They are susceptible to infection by viral, bacterial, fungal, and other pathogens. Fungi represent a particular problem for plants. There are hundreds of species of fungi that infect agriculturally important plants,[3] resulting in the loss of billions of dollars in food production each year and posing a constant threat to the world's food supply.[4] This is problematic because hundreds of millions of people worldwide do not have enough food to eat,[5] and climate change and the constantly increasing global population further exacerbate the challenge of providing sufficient food to feed the world.

The Irish Potato Famine of the mid-1800s is a prime example of the collapse of an important crop due to a pathogen epidemic, in this case infection by the fungus-like oomycete *Phytopthora infestans*.[6] Similar scares with the world banana crop have occurred more recently, caused by the pathogenic fungus *Fusarium oxysporum*.[7] The most problematic fungal and oomycete pathogens of agricultural crops are listed in Table I.1.

TABLE I.1. ***Top 10 Problematic Crop Pathogens Worldwide***

SPECIES AND COMMON/DISEASE NAMES	GROUP	MAJOR CROPS AFFECTED
Magnaporthe oryzae (rice blast)	ascomycete fungus	rice, wheat lineages
Botrytis cinerea (gray mold)	ascomycete fungus	>200 dicots: fruits, vegetables
Puccinia graminis (stem rust)	basidiomycete fungus	wheat, barley
Fusarium graminearum (Fusarium head blight)	ascomycete fungus	wheat, barley, maize
Fusarium oxysporum (Fusarium wilt)	ascomycete fungus	bananas, tomato, cotton
Blumeria graminis (powdery mildew)	ascomycete fungus	cereals
Zymoseptoria tritici (Septoria leaf blotch)	ascomycete fungus	wheat
Colletotrichum species (anthracnose)	ascomycete fungus	fruit crops
Phytophthora infestans (potato late blight)	oomycete	potato, tomato
Phakopsora pachyrhizi (Asian soybean rust)	basidiomycete fungus	soybeans

Notes: R. Dean, J.A. Van Kan, Z. A. Pretorius, K. E. Hammond-Kosack, A. Di Pietro, P. D. Spanu, J. J. Rudd, M. Dickman, R. Kahmann, J. Ellis, & G. D. Foster (2012), The top 10 fungal pathogens in molecular plant pathology, *Mol Plant Pathol 13*(4), 414–430, https://doi.org/10.1111/j.1364-3703.2011.00783.x; and S. Savary, L. Willocquet, S. J. Pethybridge, P. Esker, N. McRoberts, & A. Nelson (2019), The global burden of pathogens and pests on major food crops, *Nature Ecology & Evolution 3*(3), 430–439, https://doi.org/10.1038/s41559-018-0793-y.

Chemical pesticides for protecting crops from fungal pathogens do exist. However, many species of fungal pathogens are naturally resistant to some fungicides or readily develop resistance over time, rendering the existing treatments less and less effective.[6] For some problematic pathogens there are no effective pesticide treatments. Thus, there are global economy and health incentives to develop new fungicidal agents with novel targets and mechanisms of action to enhance protection of the world's food supply. Genetics and biochemistry are useful tools for identifying and validating new targets for any kind of pesticide or drug development. You will undertake a project this semester that explores the characterization and biochemical validation of a potential new antifungal target in crop pathogens.

PROJECT STARTING POINT

Your project is based on real work that has been published on a potential antifungal pesticide target that will be summarized below. However, for the sake of effective course design your starting point is fictional. In

Week 1 you will use bioinformatic tools to identify what this target is from its gene sequence and find information about it from the literature. Don't look ahead in this book if you don't want to spoil the surprise for the Week 1 activities. For now, we will call the gene that encodes this protein *Gene X* and will start with the fictional knowledge that *Gene X* was recently identified in a genetic screen as an essential virulence factor in *F. oxysporum*. A genetic screen relies on some method for generating a large collection of gene mutations and then testing them in some high throughput way for a phenotype of interest—in this case the inability to establish infections in a host plant. The term "virulence" refers to the ability to cause disease.

The fact that the protein product of *Gene X* is required for virulence of *F. oxysporum* makes it a potential antifungal pesticide (fungicide) target. That is, if a chemical compound could be created that specifically binds and inhibits the function of Protein X, that compound could potentially protect plants from becoming infected by fungal pathogens such as *F. oxysporum*. There are a couple more key pieces of information on *Gene X* that are relevant to your project. First, the group that identified *Gene X* in the *F. oxysporum* screen also found that *Gene X* is highly conserved across the fungal kingdom. This means that it is likely important in many if not all fungal pathogens and that a chemical inhibitor might therefore work against many fungal diseases of plants. Moreover, these scientists found that higher plants do not contain any homologs of *Gene X*. Homologs are genes in different species that were derived from a common ancestral gene during evolution. Since *Gene X* was lost from the plant lineage at some point in evolution, inhibitors of Protein X would be expected to not cause major problems for the plants you are trying to protect (although this would have to be verified experimentally once suitable inhibitors are created).

The problem is that no one knows what *Gene X* is or what the protein it codes for does. Right now, it is just a gene sequence. What is known about Protein X? Are there any high-resolution structures available for related proteins? Is it something that specific inhibitors could be developed for? Are those inhibitors likely to work against many different pathogen species? These are important questions to answer for an agricultural biotech company considering whether or not to pursue a new target like this.

PROJECT GOALS

1. Predict the identity and biochemical function of Protein X and learn what is known about its protein family using bioinformatics tools. You should find that Protein X has a clear predicted enzyme activity.
2. Express and purify Protein X from *E. coli* and test if it has the predicted enzyme activity and if it behaves similar to other members of its protein family.
3. Assess the conservation of Protein X's active site using structural modeling and substrate specificity assays. Conclude whether or not the active site is conserved enough that existing structures of homologs from model research organisms can be used as guides for inhibitor design.
4. Use information from steady-state kinetic analysis to design a screen for competitive small molecule inhibitors of Protein X and run this screen on a small set of candidate inhibitors.
5. Use protein structure modeling to predict how your best inhibitor interacts with the active site and make modifications to enhance its predicted affinity and specificity for Protein X.

LAB SAFETY POLICIES

This course has been intentionally designed to minimize the use of toxic and hazardous chemicals and dangerous lab procedures. There is very little opportunity for safety incidents to occur. Nonetheless, standard lab safety procedures enforced by the university and the department must be adhered to. This further minimizes the chances of students being injured in the lab.

Clothing requirements. Open-toe shoes such as sandals and slides are unacceptable. Footwear must entirely cover the feet. Long pants are required (no shorts, skirts, or anything that leaves bare skin exposed on the legs). Tight-fitting pants such as yoga pants are discouraged because any chemical spills on these will rapidly contact the underlying skin. Excessively loose or baggy clothing should also be avoided. Long hair must be tied back while in the lab.

Food and drink policy. No food or drink is allowed in the lab area at any time. Students should not bring food into the lab under any circumstance. If you do, it must remain stowed inside a backpack or bag. A water bottle can be stored in a backpack or bag but must be taken outside the lab (after removing all PPE) for consumption.

Personal protective equipment (PPE). PPE must be worn at all times while wet lab activities are ongoing. PPE includes lab coat, disposable lab gloves, and protective eyewear. Gloves and eyewear will be provided to students (students may bring their own safety goggles if desired). Students are expected to provide their own lab coats. A few extras will be available for students who forget to bring theirs. Gloves must be removed anytime a student exits the lab.

Other safety requirements. All personal items, such as backpacks and bags, must be kept off benchtops so they do not become contaminated with lab chemicals or interfere with work and should be kept away from exposed floor areas where they could be a tripping hazard to students walking around the lab. Dedicated areas for hanging bags, coats, and other personal items are provided and should be used. Alternatively, bags and other items may be stored completely under the work benches if there is sufficient space.

It is recommended that students wash their hands prior to leaving the lab.

Waste disposal. Any chemically contaminated disposable supplies (micropipet tips, microfuge tubes, microplates, transfer pipets, conical tubes, etc.) must be disposed of in dedicated waste containers. Do not throw these items in the regular trash. A dedicated container for biohazardous waste will be used for culture tubes or other items that come in contact with the bacterial cultures. There is a recycling bin for used gloves.

Other safety information. The lab is equipped with an eyewash station at the sink and a safety shower. The exact locations will be noted by the instructor at the start of the course. The syllabus contains links to general university policies on safety, emergency procedures, building evacuation plans, etc. Students are responsible for reviewing this information.

TIPS FOR SUCCESS

1. Keep a detailed electronic lab notebook throughout the semester. You will be writing a final report in the format of a scientific research paper at the end and will need to refer back to lots of information about your methods, data, calculations and interpretations to assemble this report. Save copies of all data and analysis files in a secure and easily

accessible location. Make sure files are named clearly. Follow the simple rule that notebooks should be sufficiently detailed for a similarly trained researcher to be able to use the information to repeat your experiments and reproduce your results at some point in the future.

2. On a similar note, make sure you carefully label everything that is yours in the lab, especially sample tubes that are stored for future use, as multiple groups working side by side in the lab can be confusing. You will need to go back and use samples from previous weeks on occasion.
3. Preparation is key! You are expected to carefully read the background information and lab instructions in each chapter ahead of time. This will help your lab group be efficient and avoid silly mistakes that could ruin experiments and cause you to stay late in the lab. In science, we only get credit for experiments that work. Part of your lab data grade each week will be based on the quality of your data and success of your experiments.
4. While it is useful to divide up work among your team members to complete experiments efficiently, remember that you are responsible for knowing how to do everything yourself. For example, you must know how to do all calculations and data analysis steps, as these will be covered in the exams.
5. The teaching assistant and the instructor are there to help you learn. Don't be afraid to ask questions to facilitate your understanding of concepts. Also, when it comes to using equipment, if you are unsure about anything, ask first!
6. Use your biochemistry textbook for additional help. Much of what you will learn in this course is covered in the Berg et al. textbook *Biochemistry*.[8]

1. PROTEIN BIOINFORMATICS

BACKGROUND AND THEORY

In the past few decades the quantity of data describing biological systems has skyrocketed due mainly to the development of high-throughput omics methods for genome (e.g., DNA sequencing), transcriptome (e.g., mRNA microarrays, RNA-seq), and proteome (e.g., protein mass spectrometry) analysis. The entire genome sequence of thousands of organisms has been determined. Today, sequencing new genomes is becoming routine. Large databases of gene and protein sequences, protein posttranslational modifications, and global gene expression data have been established for use by the scientific community. Bioinformatics is a field whose purpose is to analyze and develop the appropriate methods to analyze this vast collection of information.

The primary tool for bioinformatics is the computer, and computer programming and statistics are essential skills for the bioinformatics field. Many high-throughput experimental technologies, such as proteomics and genomics, require sophisticated computer software and statistical analyses to extract information from the massive volume of raw data. The extracted information must be organized and cataloged in a way that is accessible, useful, and easy to understand for the general scientific community. This is the job of data curators. In the lab today you are simply the users of some common and relatively simple bioinformatics tools that have been made publicly available. While many bioinformatics tools exist to analyze DNA and RNA sequence data, we are focused primarily on proteins in this course. Let's summarize some common bioinformatics applications related to proteins.

SCANNING SEQUENCES TO FIND IMPORTANT ELEMENTS, DOMAINS, AND MOTIFS

Where do genes start and stop in a genome? Where are the regulatory elements for genes? What conserved functional and structural domains exist in protein sequences? What predicted modification sites (e.g., phosphorylation) exist in proteins, and what might be the enzymes responsible for their modification? These bioinformatic functions rely mostly on text pattern searching algorithms (with the text being gene or protein sequences).

SEQUENCE SIMILARITY SEARCHING

These tools are commonly used to identify homologous genes and proteins (homologs) in other species. Homologs are genes that have evolved from a common ancestral gene. Algorithms try to align sequences and assign scores for nucleotide and amino acid identity or similarity throughout the sequences. Highly similar sequences are likely to be homologous, that is, related through evolution. There are multiple types of homologs. For example, orthologs are genes/proteins derived from a common ancestral gene/protein but that now exist in separate species. Conversely, paralogs are genes/proteins derived from a common ancestral gene/protein that exist in the same species and likely arose through gene duplication.

Much of what you will do this week involves sequence similarity searching. The basic local alignment search tool (BLAST), is one of the most common sequence similarity search algorithms and is freely available from the National Center for Biotechnology Information (NCBI), part of the National Institutes of Health's National Library of Medicine. How does BLAST work?

1. BLAST starts by using a sliding window of three amino acids from the user-provided query sequence and looks for matches in a protein sequence database, for example from the organism(s) specified by the user.
2. Whenever a match is found, the algorithm attempts to extend the similarity on either side of the match.
3. A scoring matrix is used to assign a score to each match (rare amino acids count more than common ones, similar amino acids receive lower scores than identical ones, gaps result in scoring penalties, etc.)
4. A final best alignment is built and an overall score is assigned for each match between the query sequence and a sequence in the database. A ranked list of the top scores is returned to the user.
5. An E-value is calculated, representing the probability that the final alignment could have occurred by chance (i.e., without the two proteins being related and derived from a common ancestral protein).

For more information on how BLAST works, please see the information guide provided by the NCBI at http://www.ncbi.nlm.nih.gov/books/NBK153387/. The lecture for this week has additional information as well.

PHYLOGENY

Phylogenetic tools consider similarity in gene and protein sequences on a broader scale to define evolutionary relationships between multiple species, for example defining when species split from each other on the evolutionary tree of life. Multiple sequence alignments, which you will explore today, are essential for defining phylogenetic relationships. They compare a set of sequences and align them to maximize overall similarity using scoring principles related to those in BLAST but optimized for multiple sequences rather than pairwise comparisons.

STRUCTURE PREDICTION

Prior to 2021, predicting the three-dimensional structure of proteins was very difficult. Bioinformatics tools could do a reasonably good job predicting things such as the secondary structural elements in proteins (α-helices and β-sheets) and regions of intrinsic disorder and identifying conserved structural domains present in proteins when experimentally solved structures from related proteins exist. But accurately predicting the folded structures of entire proteins was not possible. However, this all changed in 2021 with the introduction of AlphaFold,[1] an algorithm developed by Google DeepMind that has essentially solved the problem of protein folding prediction for individual proteins. Now, a predicted structure for almost every known protein sequence exists in the AlphaFold database (https://alphafold.ebi.ac.uk/). Importantly, AlphaFold does not need to use existing structures as a guide or template; it can solve new structures de novo based on large sequence alignments across many species. You will learn more about AlphaFold and get to use it later this semester.

FUNCTION PREDICTION

Proteins with similar sequences and structures often perform similar functions or have similar enzymatic activities and biochemical properties. It is therefore possible to predict protein function based on sequence similarity with other characterized proteins.

FINDING BIOINFORMATICS TOOLS

There are many valuable bioinformatics resources available for free on the internet. One useful collection of tools can be found at the NCBI website (www.ncbi.nlm.nih.gov). You will use several tools from the NCBI site, including BLAST sequence similarity searching and the PubMed literature database to predict what your protein is and does. From this site one has access to all kinds of data, including genome and proteome sequences, and many other tools for performing bioinformatics analyses on these sequences. Other useful collections of tools can be found at Expasy (http://www.expasy.org) and EMBL-EBI (https://www.ebi.ac.uk/). There are numerous other resources for bioinformatics, including many websites that provide tools and information specific to individual organisms such as the Saccharomyces Genome Database (www.yeastgenome.org), FlyBase (https://flybase.org/), the Arabidopsis Information Resource (https://www.arabidopsis.org/), and the Human Protein Atlas (https://www.proteinatlas.org/).

OBJECTIVES

Summary of problem. As outlined in the Introduction, your projects start with the knowledge that genetic screening has identified a gene (*Gene X*) that is required for plant infection by the fungal pathogen *Fusarium oxysporum*. This fact makes it an attractive potential target for new fungicides. However, you know nothing about the function of the protein that it encodes. Developing chemical pesticides that target this protein will first require knowledge about how this protein functions. The overall goal for this semester is to purify this protein and characterize some of its biochemical properties to determine if it could be a useful antifungal pesticide target.

This week you are being provided with the DNA sequence of the putative coding region of *Gene X* from *F. oxysporum*. You will explore some common protein bioinformatic tools and literature searching tools to learn about the protein encoded by *Gene X*. Specifically, your primary objectives for today are to

1. Generate the amino acid sequence of Protein X from its DNA coding sequence,
2. Identify homologs of Protein X in common research organisms using BLAST,
3. Identify any conserved functional domains in Protein X,
4. Perform a multiple sequence alignment between Protein X and the homologs identified in step 2 to determine the most conserved sequence regions,
5. Learn about the functions of Protein X homologs using literature searching tools,

6. Generate a hypothesis about the enzyme activity that Protein X possesses based on your findings, and
7. Identify a mutation predicted to inactivate Protein X and design primers to create that mutation in *Gene X*.

LAB INSTRUCTIONS

PART 1. BIOINFORMATICS EXERCISES

1. Expasy Translate tool
   A. Navigate to the Expasy bioinformatics portal at http://www.expasy.org/. Check the box for the "Proteins & Proteomes" category at left and then find the "Translate" tool button under "Other Resources of SIB Groups." On the next page, click "Browse the resource website" to open the translate tool. This tool converts a gene coding sequence into its corresponding protein sequence based on the known codon:amino acid relationships.
   B. Copy the coding sequence for your target gene (provided to your group as a .txt text file) and paste it into the query window. Choose the "Compact" output format, check just the "forward" box under DNA strands, then hit the "TRANSLATE!" button.
   C. Copy the resulting amino acid sequence encoded by your target gene and save it to your group's designated data folder as a text file. You will need the sequence again later today and later in the course.
2. BLAST sequence similarity search
   Now your goal is to determine if there are closely related proteins in other species that might give you clues to the identity and function of Protein X. BLAST is a great tool for this purpose.
   A. Go to www.ncbi.nlm.nih.gov and follow the link from the main page to the BLAST program and then select "Protein BLAST."
   B. Copy the protein sequence you saved from the translate tool in step 1.
   C. Paste the Protein X sequence into the query window. Make sure that the recommended "ClusteredNR database" radio button is selected under the "Database" section. Keep other settings on default selections but add *Saccharomyces cerevisiae*, *Schizosaccharomyces pombe*, and *Homo sapiens* to the "Organism" section. You can select all three at the same time to search against using the "Add Organism" button after finding each.
   NOTE: These species are heavily studied, so if they have proteins related to Protein X it is more likely that something will already be known about them than in other species.
   D. Save the results of your search (screenshots are fine), particularly info on the top matches from each species. Assess the E-value for the top score. Clear homologs often have E-value scores below 1e-25. The more negative the E-value, the more similar the sequences are and the more confident you can be that they are derived from a common ancestral protein. Make sure you have a match from each of the three species. This should provide you with names that you can now use to look up articles published on this protein family in PubMed and/or Google Scholar in step 5 below. Was there a common name you found for the homologs in all three species?
   E. In addition to screenshots, you will also need to find and save the protein sequences for the top similarity hits from the three species as text files, just like you did for your own Protein X sequence. To find the sequences, click on the hit name in the Cluster Composition column, then its Accession link, and finally the FASTA link. Copy the entire sequences, with FASTA header lines, and save them in a text file.
3. Multiple sequence alignment
   The BLAST search should have identified the most closely related protein sequences from three diverse eukaryotes (fission yeast, budding yeast, and humans). These represent putative orthologs from these species that might have similar functions and properties because they are derived from a common ancestral protein through evolution. Now,

the question is, what do all of these proteins have in common (in their amino acid sequence) and can you use this to help learn about the possible functions and properties of your protein? One way to do this is with a multiple sequence alignment. Instead of comparing two sequences as BLAST does, you can compare a larger number of sequences to discover what they have in common using tools such as Clustal Omega. This exercise is also useful for finding the regions of your protein that are likely to be the most functionally important (these will tend to be most resistant to evolutionary change and therefore will be most similar across species).

A. Access Clustal Omega (ideally using Chrome) through the www.expasy.org site (filter using the "Proteins & Proteomes" checkbox again, and then select the one labeled "Uniprot ClustalO.)" Click "Browse the resource website" on the next page.
B. Once on the Clustal Omega page, you need to paste the FASTA sequences of Protein X and the three putative orthologs you found from BLAST into the window at the top (so, four sequences in total). Remember that these are saved from above as text files. Simplify the FASTA header line with just a species abbreviation (e.g., S_cer, S_pom, H_sap); the FASTA header line is the text following the > symbol. Click on the "Align Sequences" button at the bottom. Please note that you must have a correct FASTA header line for each sequence to make the alignment work (> followed by a text string).
C. When the alignment is finished you will be able to click on the "Completed" link. On the results page, click on the Download button and select the first format option (Clustal format with numbers). You can then highlight and copy the alignment text and paste it into Notepad. Save this file to your group's data storage folder as well. If using a Firefox browser you may have to save the file to your computer first, open it in Word or Notepad, select and copy all text, and then paste it into a new plain text document to get the alignment to look correct.

4. Conserved domain search

Another way to get clues to possible functions for an uncharacterized protein is to look for conserved domains. This is similar to BLAST, but instead of looking for highly similar proteins based on the overall sequence, the algorithm will look for smaller regions (domains) that are commonly found in other proteins and have characterized structures and/or functions. It searches your query sequence against a dedicated database of characterized protein domains.

A. Go back to the NCBI home page (www.ncbi.nlm.nih.gov) and select "Domains and Structures" from the left-hand menu.
B. Use your Protein X sequence to search for conserved domains using the CD Search tool. Default settings should work fine.
C. What does this result tell you about likely functions or biochemical properties of your protein? Does it suggest that your protein might be an enzyme that catalyzes a particular chemical reaction?
D. Use the "Conserved Domains" tab to find information about how these algorithms identify conserved protein domains and what information can be accessed in the output.
E. Save the output from the CD Search as a screenshot image and record a text description of your findings in your notebook for later reference.

5. PubMed literature search

A. Use PubMed (also found from NCBI main page) and Google Scholar searches to look for articles on orthologs of Protein X to learn what is known about it. You may want to use the advanced search function in PubMed and limit your search to title and/or abstract words.
B. Find and save two review articles that discuss orthologs of Protein X in other species and two primary research articles describing biological functions for orthologs of Protein X. (You can search specifically for reviews by selecting "Review" as publication type in the PubMed search criteria and by clicking "Review Articles" on the

left-side menu in Google Scholar). HINT—use the common name for Protein X that you discovered from the BLAST results when searching for articles.

6. Additional biochemical and biophysical properties
   Expasy has links to many great bioinformatic tools. Scan through the descriptions of the different tools linked on the main page to get a sense of what types of bioinformatic applications are available here.
   Using the "Proteins & Proteomes" filter again, find and open the ProtParam tool. Paste in your Protein X amino acid sequence, run the tool, and copy and save the output information as a text file again. It contains some useful information such as amino acid content, isoelectric point, exact molecular mass, extinction coefficient, and other physicochemical properties.
7. Finding and viewing high resolution protein structures
   A. We will now see if there are any known high-resolution atomic structures for related proteins that might give you an idea of what your protein looks like. Navigate to the Protein Data Bank (PDB) at www.rcsb.org and type the common name for your protein family deduced from your BLAST and PubMed exercises above in the search window at top. You should see a list of possible matches appear as you type.
   B. Explore the different matches that appear and note how many different structures are available for members of this protein family. You should find structures for both human and *S. cerevisiae* homologs bound to substrate molecules. Look for one annotated "budding yeast Cdc14p (C283S) bound to a Swi6 phosphopeptide." Click on the title and then use the "Download Files" dropdown menu to download the atomic coordinates file in PDBx/mmCIF format. Record the PDB ID for this structure. You will need to use this structure later in the course. If you don't find any structures, let the teaching assistant or the instructor know so they can check how you are searching for structures.
   C. Now, go to the AlphaFold protein structure database (https://alphafold.ebi.ac.uk/). This is a database of protein structures predicted by Deep Mind's AlphaFold algorithm. Has AlphaFold already predicted the atomic structure of your *Fusarium oxysporum* protein? If so, download the atomic coordinates file for this structure and save it as well. You may have to first go to https://www.uniprot.org/ and search for your protein and use the "Primary accession" from Uniprot to search the AlphaFold database.

PART 2. PRIMER DESIGN FOR SITE-DIRECTED MUTAGENESIS

At this point your group should be able to state a hypothesis regarding the identity and specific enzyme activity of your target protein based on its similarity to other characterized proteins. And you should be able to give Protein X a more appropriate name now. Using what you have learned and found from the conserved domain search, sequence alignments, structures, and literature searches, see if you can locate your protein's predicted enzyme active site. This family of enzymes has a single amino acid in the active site that plays an essential role in initiating catalysis. You must know the location of the active site for this next exercise.

1. Open the SnapGene Viewer software on your computer and create a new sequence by pasting in the coding sequence of *Gene X*. Then, select the entire sequence (Ctrl + A) and click on the "Show Translations" icon on the left menu column. Choose "+1 Frame Only." This does the same thing as the Expasy Translate tool. Now you can see your protein's sequence together with the gene coding sequence. Save this SnapGene file in your group's data storage folder for future reference.

2. Find the active site and the predicted catalytic amino acid of your enzyme in the SnapGene protein sequence. Refer to the publication that describes the structure you downloaded from the PDB in Part 1, step 7 for guidance in finding this amino acid (hint: they used an inactive mutant version of the enzyme where this amino acid was altered to get their structure). Highlight this amino acid in your SnapGene file and use the Add Feature tool on the Features menu to label this amino acid as the catalytic site.
3. Next, find a codon table online and choose a codon to replace the catalytic amino acid's codon. You want to create the same catalytically inactive mutant used in the structure papers.
4. You will use a method called inverse PCR to generate these mutants in lab next week. PCR requires two synthetic single-stranded oligonucleotide primers to amplify a sequence of interest. Mutagenesis can be accomplished by inverse PCR if the desired mutations are incorporated into the PCR primers, which can be conveniently ordered commercially. You will learn more about the mutagenesis method next week, but for now we will use the primer design tool from Takara Bio found at https://www.takarabio.com/learning-centers/cloning/primer-design-and-other-tools.
   A. Click the "Launch" link in the "Design your primers" section.
   B. Choose "Mutagenesis" in Box 1.
   C. Paste the *Gene X* coding sequence into the window in Box 2.
   D. Find and highlight the codon for the catalytic amino acid in the sequence that appears on the right.
   E. In Box 3, select "Replace Selection" and enter the codon for the new amino acid you chose to create the inactive mutant.
   F. Click "Design Primers" in Box 4.
   G. You will see the primer sequences appear. Click "Download Results" at the bottom to obtain a .csv file with your mutagenic primer sequences.
5. The reasons for doing this will become clearer next week when we cover the topic of structure-function experiments and begin discussing the process of purifying proteins for biochemical characterization. You will use these primers next week to actually introduce the mutation into the coding sequence for Protein X.

DATA ANALYSIS AND SUBMISSION

Your group needs to submit the following files generated during today's lab exercises:

1. Text file of the output from the Translate tool—your protein's predicted amino acid sequence.
2. Screenshot of your BLAST search results, clearly showing the top hits from each species and their E-values.
3. Text file with FASTA sequences of the top hit from each species in the BLAST search.
4. The text file output from your Clustal Omega multiple sequence alignment.
5. Screenshots of the output from your Conserved Domain search.
6. Four PDF files from your PubMed literature searches.
7. Text output file from the ProtParam analysis of your protein sequence.
8. The PDB coordinates file of the crystal structure of the *S. cerevisiae* Protein X homolog bound to a substrate.

In addition to the data files, your group must also submit a Word document with the following information:

1. A hypothesis statement regarding the identity (common name) and specific enzymatic activity of Protein X (i.e., what specific chemical reaction is it predicted to catalyze?).

2. The identity (names) of any conserved domains found in your protein with a description of their known functions/properties.
3. The location of the putative active site in your protein (the best way to do this is to highlight it within the full amino acid sequence), with the amino acid(s) being altered by site-directed mutagenesis underlined.
4. The sequence of the primers you designed using the Takara Bio design tool to alter your protein's putative active site. Clearly indicate where the changes are from the original (wild-type) sequence.

2. PROTEIN STRUCTURE-FUNCTION ANALYSIS

BACKGROUND AND THEORY

BRIEF REVIEW OF PROTEIN STRUCTURE

You will spend a significant amount of time in this course looking at protein sequences and protein 3D structures. While you should be quite familiar with the basics of protein structure at this point, it makes sense to review the fundamentals again here.

All proteins are composed of linear chains of 20 amino acids in orders defined by their gene sequences. The order of amino acids in the polymeric polypeptide chains is called a protein's primary structure (Figure 2.1). It is important to understand that polypeptide chains are repeating units of three atoms derived from each amino

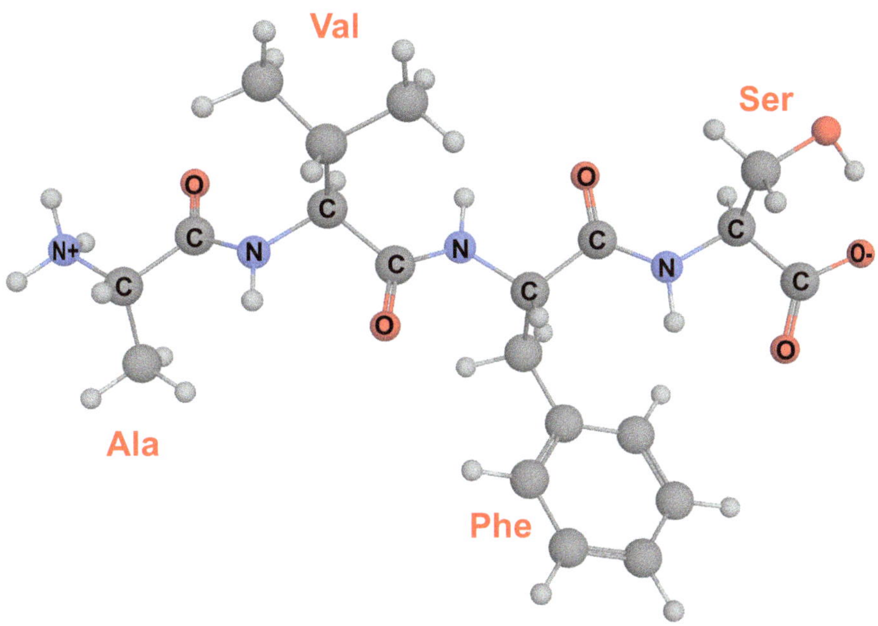

FIGURE 2.1. Protein primary structure. A short protein region consisting of four amino acids in the sequence alanine (Ala)—valine (Val)—phenylalanine (Phe)—serine (Ser). Atoms are nitrogen (N) in blue, oxygen (O) in red, carbon (C) in large gray, and hydrogen in small gray. Notice the backbone repeating units of N-C-C. The bond linking adjacent amino acids is called the peptide bond. Image generated with Molecular Operating Environment.

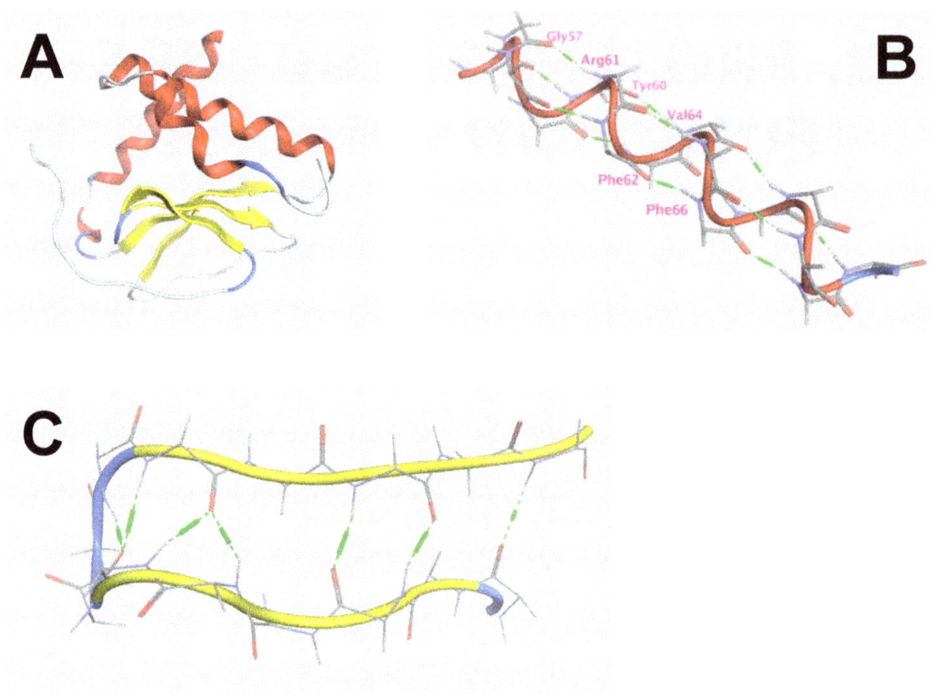

FIGURE 2.2. Protein secondary structure. (A) the common secondary structures, alpha helices and beta strands, shown in red and yellow, respectively, in an example protein structure. Turns (blue) and random coil regions (white) are also visible. (B) A right-handed alpha helix is created by hydrogen bonds (green) between the backbone N-H and C=O groups of every fourth amino acid. (C) an example of antiparallel beta strands stabilized by backbone hydrogen bonds (green). Images generated with Molecular Operating Environment.

acid: a nitrogen (N), a carbon linked to the variable side chain R group (Cα), and a carbon double-bonded to an oxygen (C=O).

The polypeptide chain can adopt several local secondary structures depending on the R groups of the constituent amino acids. The two common secondary structures are alpha helices and beta sheets (Figure 2.2). Alpha helices are typically formed from hydrogen bonding between the N-H and C=O backbone groups of amino acids spaced four positions apart, with the backbone twisting in a right-handed direction. Hydrogen bonds are a type of noncovalent bond between hydrogens and electronegative atoms (e.g., oxygen) that play a central role in molecular interactions in biology (covered in more detail in Chapter 11). Beta sheets are also formed by hydrogen bonding between backbone N-H and C=O groups and can involve peptide segments running parallel or antiparallel to each other. Parts of a polypeptide chain can also lack secondary structure, and these are typically called random coils or disordered regions.

Tertiary structure refers to the overall 3D arrangement of atoms in a protein, resulting from the packing of the various secondary structure elements into a defined arrangement (Figure 2.3A). The determinants of protein folding into 3D tertiary structure are complex. A general rule is that the core of folded proteins consists of hydrophobic amino acids and polypeptide regions, whereas surfaces consist mostly of polar amino acids where there is exposure to solvent. Quaternary structure (Figure 2.3B) exists for some proteins that are

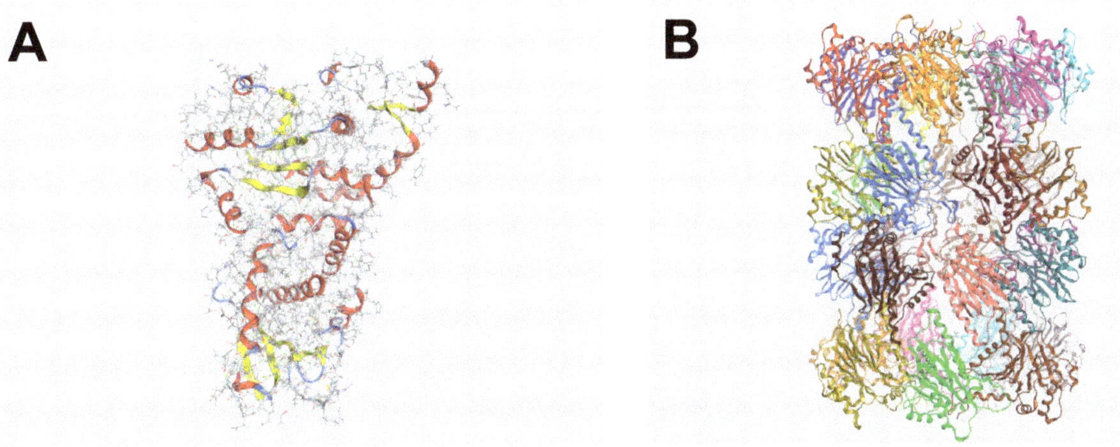

FIGURE 2.3. Protein tertiary and quaternary structure. (A) The final packing of amino acids and secondary structures into a 3D shape is referred to as protein tertiary structure. Here the secondary structure elements are shown as in Figure 2.2, and each amino acid is shown in stick format with coloring as in Figure 2.1, created from RCSB Protein Data Bank (PDB) ID: 5XW5, a protein structure you will be working with during this course. (B) Many proteins function as subunits of stable multisubunit complexes, referred to as quaternary structure. Here, the 20S proteasome complex is used as an example, with each individual protein subunit colored uniquely in secondary structure depiction (RCSB PDB ID: 3C91). The proteasome is used by eukaryotic organisms to degrade proteins and recycle amino acids. Images generated with Molecular Operating Environment.

multimeric through either stable interactions with identical protein molecules (homooligomers) or different protein molecules (heterooligomers). Finally, proteins can be modified by a wide variety of chemical functional groups, usually through the side chains of polar amino acids, after they have been translated and folded. These so-called posttranslational modifications can be considered a fifth level of protein structure. They are typically dynamic and reversible and play important roles in regulating protein function. You will learn about one of the most common regulatory posttranslational modifications, phosphorylation, during this course.

The different side chain R groups of the amino acids in proteins are central to dictating not only a protein's structure but also its functional properties. The side chains are involved in mediating molecular interactions, driving the chemistry of enzyme-catalyzed reactions, and serving as targets for protein regulation, among other things. It is important for you to know the 20 amino acids and the properties of their side chain groups (Figure 2.4).

One of the repeating polypeptide backbone bonds has a very important property that greatly influences protein structure. The bond between the C=O carbon and the N of the next amino acid in the chain is called the peptide bond. It is the bond that is formed to link amino acids together through a condensation reaction (loss of water) catalyzed by the ribosome during translation of an mRNA into a protein. The peptide bond exhibits partial double-bond character due to the phenomenon of resonance with the C=O bond. As a result of its partial double-bond character, the peptide bond is planar (all four atoms C, O, N, H lie in a single plane) and rigid (no rotation of atoms relative to each other). This feature of the peptide bond creates constraints on how a polypeptide chain can fold. Importantly, it limits the possible rotational angles of the Cα-C and N-Cα single bonds before and after the peptide bond (Figure 2.5). There are some nice YouTube videos that explain the planarity of the peptide bond and its implications for protein structure if you are confused by this or would like to learn more.

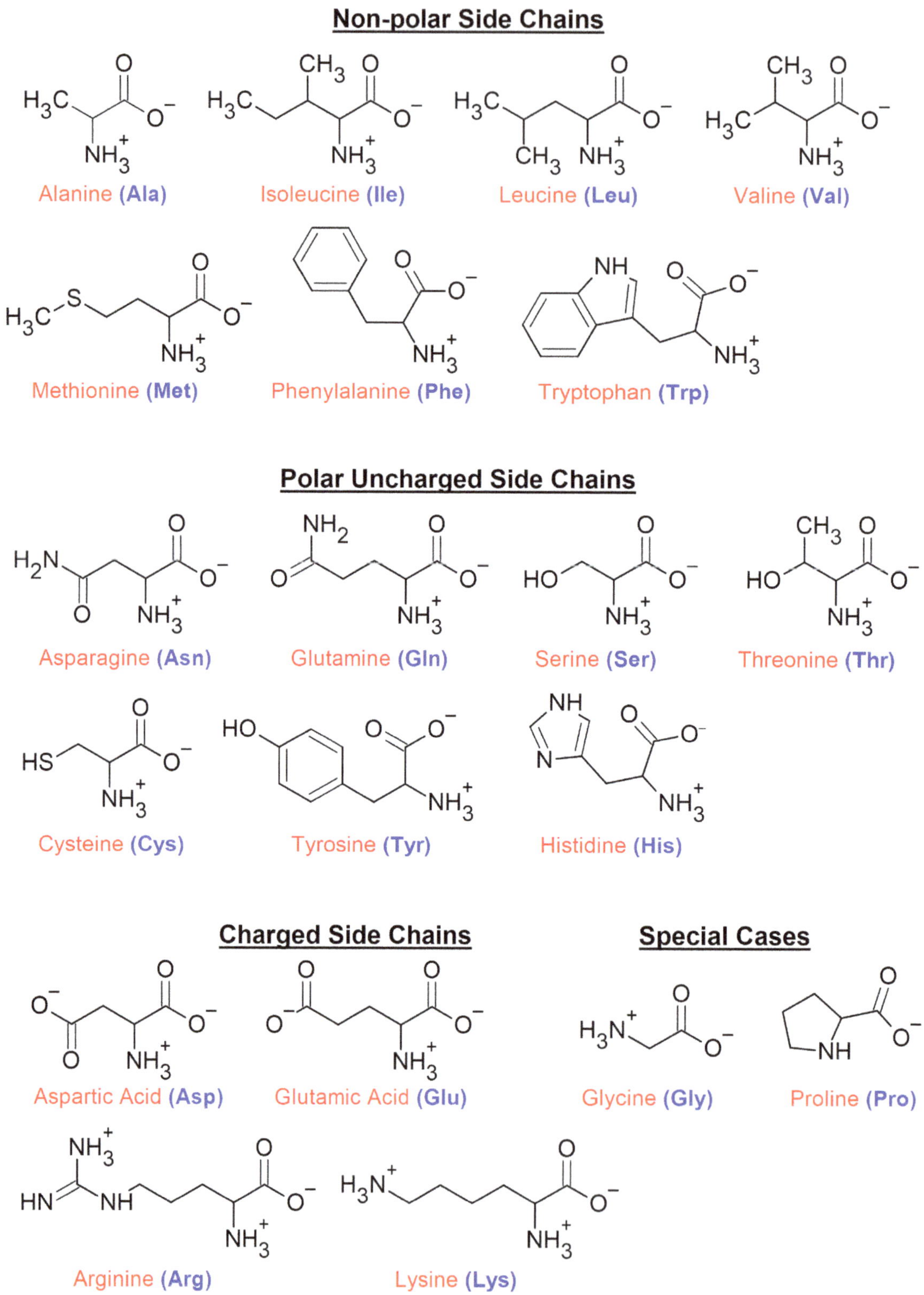

FIGURE 2.4. Amino acid building blocks of proteins. The 20 genetically encoded amino acids that contribute to protein structure are shown with predominant charge states at a physiological pH of 7. They are organized based on chemical properties of the side chain functional group. Standard three-letter abbreviations are included in parentheses. Single-letter abbreviations are not shown.

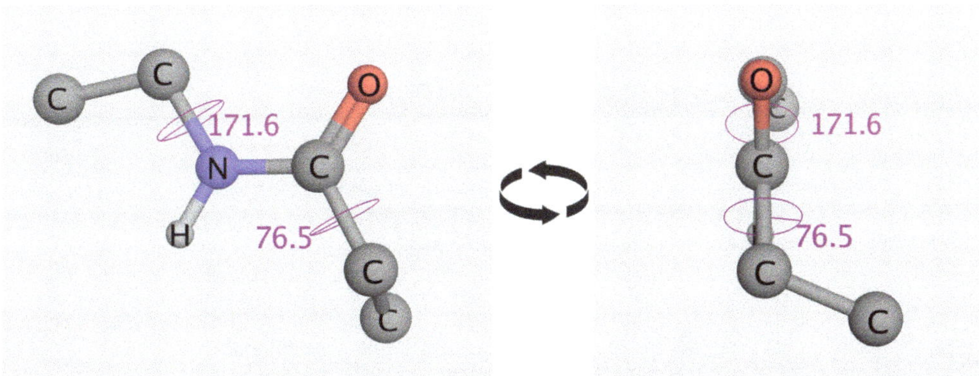

FIGURE 2.5. Peptide bonds are planar. *Left*, a single peptide bond from a protein with adjacent Cα and side chain C. The rotational angles of the bonds linking the peptide bond to the adjacent Cα are indicated in purple. *Right*, when rotated 90° the planar nature of the peptide bond is evident (N-H, C=O all lie in same plane). Images generated with Molecular Operating Environment.

THE CONCEPT OF PROTEIN STRUCTURE-FUNCTION STUDIES

One of the most important goals of biochemical research is to understand how the structure of biological molecules defines their function within biological systems. This is particularly true for proteins, which fold into an amazingly diverse array of structures and carry out an equally amazing diversity of biological functions. Protein structure is ultimately determined by the amino acid sequence. Altering the sequence changes protein structure, and researchers can then study how these changes affect protein function.

In the lab, it is possible to engineer mutations (or sequence changes) in the genes that encode for proteins. The mutant proteins can be characterized and compared to the original wild-type protein to determine the functional importance of individual amino acids, conserved amino acid motifs, and entire structural domains. We call this general approach ***structure-function analysis***. An essential component of structure-function analysis is the ability to engineer mutations in specific amino acid sequences of interest. This is the topic of today's lab exercise.

PCR-BASED GENERATION OF SITE-SPECIFIC MUTANTS

There are a variety of methods available for engineering mutations into gene sequences. Today, commercially available kits simplify the process. The majority of methods for creating small changes (individual amino acids or small sequence insertions or deletions) share some common principles. First, the gene being targeted must be cloned into an *E. coli*–compatible plasmid. Second, short single-stranded DNA oligonucleotides (also called primers) are designed that contain the desired codon changes (substitutions, insertions, or deletions) for altering the encoded protein's amino acid sequence. Third, polymerase chain reaction (PCR) with a thermal cycler, the primers, and a thermally stable DNA polymerase enzyme are used to synthesize new plasmid molecules that now contain the desired sequence mutations. The PCR technology is central to this mutagenesis process. Make sure you are comfortable with how PCR works at the molecular level (Figure 2.6). PCR involves repeated cycles of (1) denaturation of DNA strands, (2) annealing of the short single-stranded DNA primers to the target DNA, and (3) new DNA strand synthesis by the thermally stable DNA polymerase. In short, PCR provides rapid amplification of a specific DNA sequence whose ends are defined by the two primers.

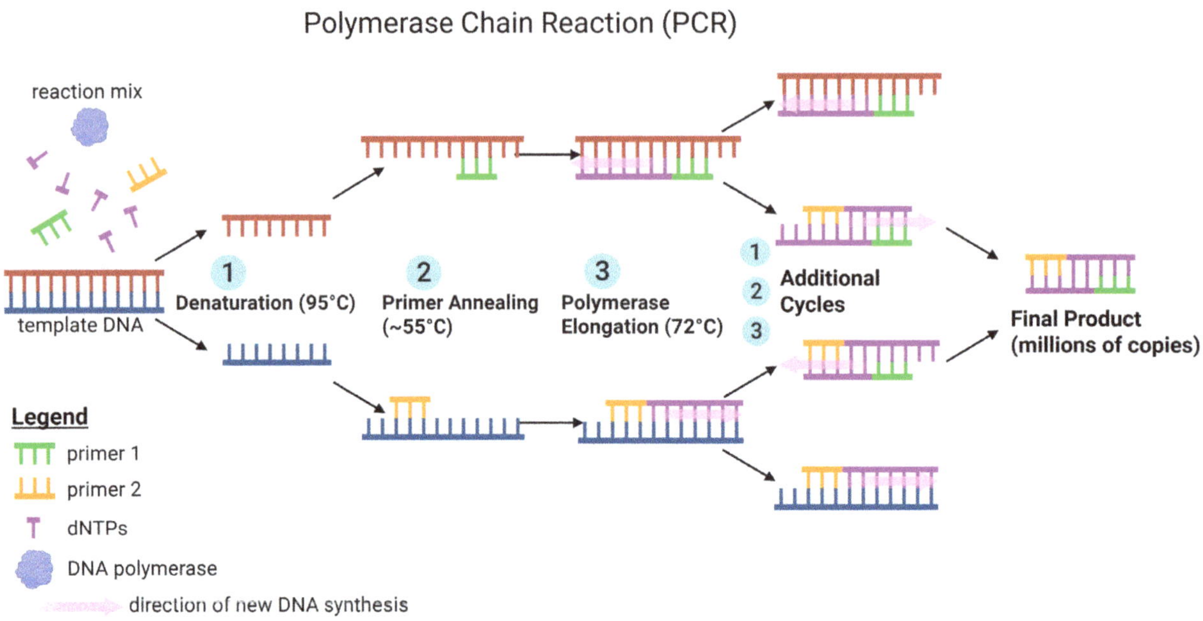

FIGURE 2.6. Overview of PCR. Polymerase chain reaction results in the specific amplification of a DNA sequence defined by two single-strand oligonucleotide primers that anneal to opposite strands of a starting template DNA molecule. The primer sequences are user-defined and determine the ends of the final amplified DNA sequence. A thermostable DNA polymerase is used to synthesize new DNA strands from the 3' end of each primer after annealing to a template DNA strand using the four deoxnucleotides (dNTPs) provided in the reaction mixture with appropriate buffer and salt conditions. Repeated cycles of (1) DNA strand denaturation at high temperature, (2) annealing of the primers to their complementary sequence at lower temperature, and (3) new strand synthesis by the thermostable polymerase leads to exponential amplification of the desired product DNA molecule. Created in BioRender. Hall, M. (2026) https://BioRender.com/58nzdhc.

Following a PCR-based mutagenesis reaction, the new mutation-containing plasmid molecules must be recovered, amplified, and separated from the original template plasmid. To accomplish this, the reactions are transformed into *E. coli* cells, selecting for antibiotic-resistant growth conferred by the plasmid. Colonies that emerge on agar media plates under antibiotic selection must then be screened to find plasmids that contain the desired sequence changes. This is commonly done by directly sequencing the plasmids after recovering them from liquid bacterial cultures started from single transformant colonies. Other methods for screening exist as well, including PCR using primers that will specifically amplify a product only if the desired mutation is present.

We will be using a simple inverse PCR method to incorporate the putative active site mutation you chose last week into the *Gene X* (from here on called *F. oxysporum CDC14*) sequence (Figure 2.7). The mutation is incorporated into partially overlapping PCR primers. PCR amplification using the mutagenic primers leads to a linear double-stranded DNA molecule whose two ends share a short stretch of identical sequence. An exonuclease is used to degrade one DNA end to generate complementary single-strand ends that will anneal to give a circular plasmid again. When transformed into *E. coli*, native DNA polymerase and ligase activities fill in the remaining gaps.

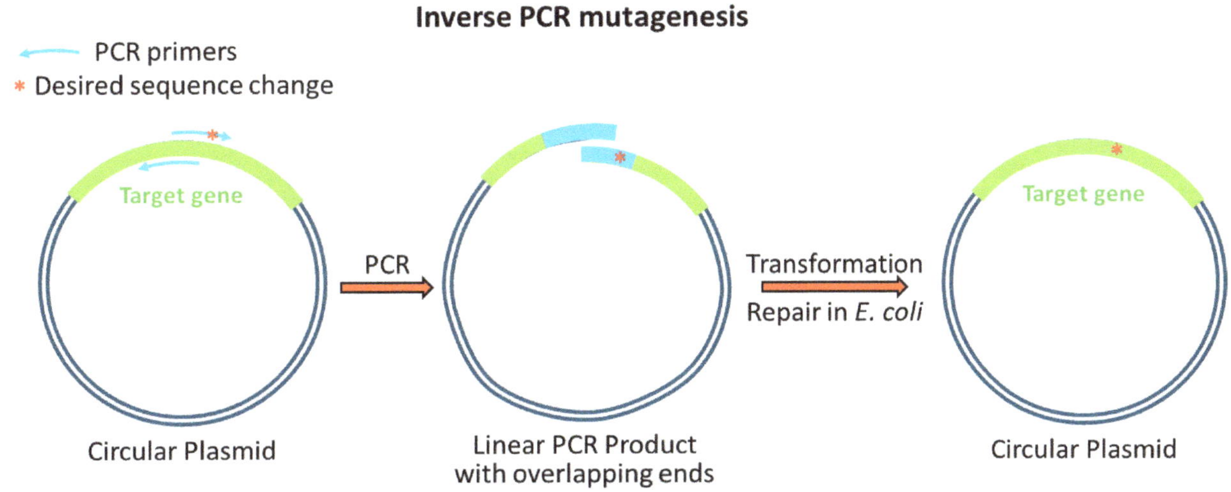

FIGURE 2.7. Mutagenesis by inverse PCR in the Takara Bio In-Fusion® system. Two PCR primers are designed to amplify the entire plasmid and have a region of ~15–30 bp of overlap at their 5' ends. The primers anneal near the desired mutation site, and the altered sequence (*) is incorporated into one or both primers. The resulting PCR amplicon is a linear version of the plasmid, with identical repeats at the ends defined by the overlapping primer sequence. After exonuclease-catalyzed end resection, annealing between the complementary repeat region creates a circular molecule. When transformed into *E. coli* cells, endogenous enzymes fully repair the circular plasmid, which differs from the original plasmid only in the sequence change engineered into the primers.

An important feature of this mutagenesis protocol is the use of the DpnI restriction endonuclease to destroy the parental wild-type plasmid DNA strands to increase the efficiency of recovering mutant molecules after the bacterial transformation. DpnI only cuts methylated DNA. Plasmids purified from *E. coli* are naturally methylated by endogenous methyltransferases, whereas the new mutant plasmid molecules synthesized in a test tube during the inverse PCR reaction will not be methylated and therefore won't be recognized and cut by DpnI.

VERIFYING SUCCESSFUL MUTAGENESIS: DNA SEQUENCING

How is DNA sequence determined? For decades, the most common method for sequencing of individual molecules was the Sanger dideoxynucleotide method, invented in 1977 by Dr. Frederick Sanger in Cambridge at the MRC Lab of Molecular Biology. It is still commonly used, now with a number of modifications to make it more automated and routine. The process depends on the use of a low concentration of chain-terminating dideoxynucleotides that lack a 3'-OH (Figure 2.8) and is summarized in Figure 2.9. A video that explains the process in animated form is available at https://www.youtube.com/watch?v=e2G5zx-OJIw.

In short, the target DNA is subjected to an in vitro DNA polymerization reaction initiated by an oligonucleotide primer and a DNA polymerase in the presence of regular deoxynucleotides (G,A,T,C) and a smaller concentration of dideoxynucleotides linked to different color fluorescent tags. When one of the dideoxynucleotides is randomly incorporated into the new DNA molecule it is a dead end, and synthesis of the DNA strand stops. This is because the dideoxynucleotide does not have a 3'-OH group on the sugar, which the polymerase requires to join to the next residue. Random incorporation of dideoxynucleotides leads to a collection

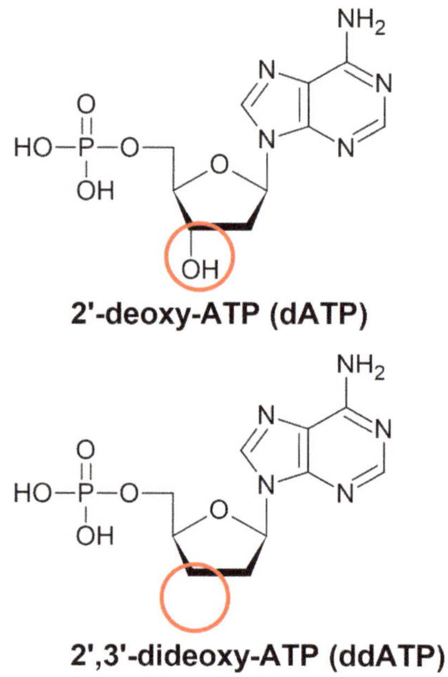

FIGURE 2.8. An example of the structure of dideoxynucleotides (ddNTPs) used as DNA synthesis chain terminators in Sanger DNA sequencing compared to the natural deoxynucleotides (dNTPs) that are building blocks of DNA. The 3'-OH in the sugar (circled in red) is essential for DNA polymerases to attach the next dNTP in the growing chain. This OH is missing in ddNTPs.

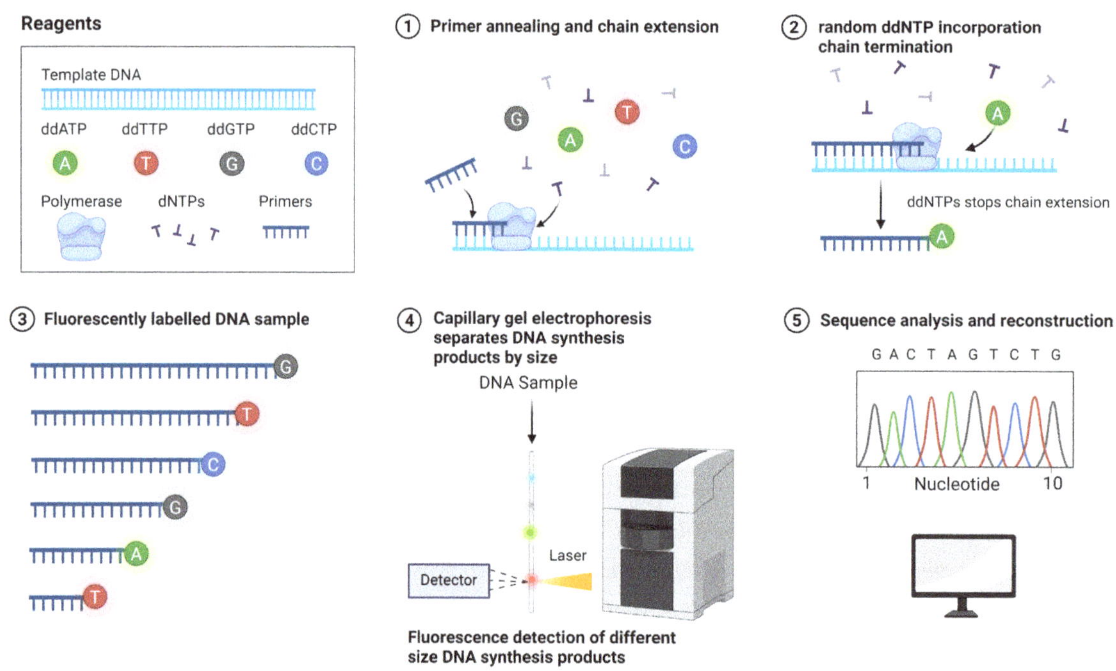

FIGURE 2.9. The overall process of ddNTP-based Sanger DNA sequencing. Importantly, the template to be sequenced is denatured, and a primer anneals ahead of the region of interest (1). DNA polymerase catalyzes new strand synthesis in the presence of a high concentration of dNTPs and low concentration of ddNTPs (2). Each ddNTP is linked to a different color fluorophore. Each newly synthesized DNA molecule will have a ddNTP incorporated essentially at random, leading to a collection of product DNA strands of differing length based on where a ddNTP was incorporated (3). The product population is separated by size using a process called capillary electrophoresis, and the color of each product is measured as it migrates by a fluorescence detector (4). The output is a chromatogram of fluorescence peaks where the sequence of colors defines the DNA sequence in the template (5). Created by Created in BioRender. Hall, M. (2026) https://BioRender.com/xa39gps.

of products of different lengths, with each fluorescing a different color depending on the identity of the base that terminated the reaction. Products are separated by length by either gel or capillary electrophoresis with a fluorescence detector to identify the color of each band as they travel by. The sequence can be directly read from the pattern of fluorescence.

While Sanger sequencing is still used for some applications, such as verifying products of gene mutagenesis reactions, so-called next-gen or massively parallel strategies have become the predominant method for obtaining DNA sequence information, including the rapid sequencing of entire genomes that can be achieved through popular commercial services such as 23andMe and AncestryDNA. While there are multiple platforms, the common features are that sequences do not need to be cloned into a plasmid first, and many DNA molecules can be sequenced simultaneously. While some still rely on PCR-like amplification, new strand synthesis, and use of dideoxynucleotide chain termination technology, others are now capable of direct sequence reading of single DNA molecules. A good example is the Oxford Nanopore technology. Details of these technologies are outside the scope of this course.

GENERAL CONCEPTS OF ANALYTICAL MEASUREMENTS

Today you will begin preparing for making accurate and precise measurements of such things as enzyme-catalyzed reaction rates. This section covers some important basic principles of analytical instruments and measurements that are important to understand and review.

All analytical measurements have a degree of error associated with them. Nothing measures with 100% accuracy. There are many sources of error in analytical measurements. Some are inherent to the instrument or measuring device. Others come from the operator or environmental conditions (temperature, humidity, etc.).

The error associated with the instrument itself can be minimized by accurate **calibration.** When the accuracy of an analytical instrument becomes less than acceptable, it must be calibrated again. In general, to calibrate an instrument a **standard** of known quantity (for example, the density of water in the lab exercise today) is measured, and the instrument is adjusted until the known value is obtained.

The **error** in an analytical measurement is defined as the difference between the measured value and the theoretical value. There are two important components of analytical measurements that contribute to error: **accuracy** and **precision.** It is important to understand the difference between them, as these terms are often confused.

Accuracy is the difference between the experimentally measured value and the true (or theoretical) value. Precision is the variation associated with repeated measurements. A measurement can be very precise but highly inaccurate if, for example, the instrument is poorly calibrated. It is much less likely for a measurement to be highly accurate but imprecise.

There are different ways to express accuracy, but one common and simple method is the percent error.

$$\% \text{ error} = (O\text{-}T)/T \times 100$$

where O is the observed value and T is the theoretical value.

Precision can also be expressed in different ways, but one common format is standard deviation, a statistical value reflecting the variability of repeated measurements relative to their average. Standard deviation is useful for determining if differences in experimental measurements are meaningful or just reflect the natural level of variation. There are many other types of statistical tests that can be performed to determine if measured differences are likely to be real. You will explore some of these later in the course. The basic equation for standard deviation is

$$\text{standard deviation } (\sigma) = \sqrt{\Sigma\,(x - \overline{x})^2/N}$$

where x is each measurement, $\overline{x}$ is the mean, and N is the total number of measurements.

OBJECTIVES

The first objective today is to create the mutation that you selected last week to inactivate your predicted enzyme activity. This mutation is very important because it can be used to (1) confirm that your enzyme acts by a similar mechanism to other characterized Cdc14 phosphatases and (2) verify that any enzyme activity you measure in your purified protein preparation in the upcoming weeks is really due to your target enzyme and not a contaminating activity from the *E. coli* cells. The mutant gene could also be used to study the function of the enzyme's activity by introducing the mutant back into its native organism (*F. oxysporum*) if the genetic and molecular biology tools for doing that are available.

In addition to running the mutagenesis reactions and transforming the products into *E. coli* cells for recovery, each group will make the media needed to grow *E. coli* cells next week for production of their protein. Finally, you will ensure that you have accurate, functional micropipettors to use this semester while reinforcing the concepts of accuracy, precision, calibration, and experimental variation.

LAB INSTRUCTIONS

PART 1. MUTAGENESIS OF THE PREDICTED CATALYTIC AMINO ACID

You have been provided with the following reagents:

- A 10 ng/µl stock of the expression plasmid containing the *F. oxysporum CDC14* gene that you will use in the upcoming weeks to purify the FoCdc14 enzyme.
- The forward and reverse PCR primers (designed last week) that contain the sequence to alter the putative catalytic Cys to Ser in FoCdc14. Each primer is at a concentration of 5 µM.
- A small tube of sterile ultrapure water.
- A PCR reaction tube.
- Two LB-ampicillin plates with sterile plating beads.
- Two sterile 14 ml culture tubes for transformations.

The following additional reagents are kept on ice by the instructor, who will provide access when needed:

- DNA polymerase (contains buffer, salts, and dNTPs)
- DpnI enzyme mixture

- In-Fusion exonuclease enzyme mixture
- Competent *E. coli* cells for transformation

1. Set up your PCR reactions in the provided PCR tube by mixing the following:
   5.5 μl H_2O
   2 μl plasmid template DNA
   2.5 μl primer 1
   2.5 μl primer 2
   12.5 μl DNA polymerase master mix
   25 μl total
2. Close the cap securely and label your tube with your group number using a marker. Place the tube in the thermal cycler. Once all groups are ready, the instructor will initiate the method. The thermal cycling method is programmed as follows and will take a little over an hour to complete:
   98°C, 1 minute
   35 cycles of:
      98°C, 10 seconds
      55°C, 5 seconds
      72°C, 1 minute 15 seconds (10 seconds/kb, plasmid is 7.1 kb)
   72°C, 3 minutes
   4°C hold
3. Mix 2 μl of DpnI enzyme mix with 10 μl of the PCR reaction after it has completed. Incubate at 37°C for 15 minutes and then 80°C for 10 minutes.
4. Mix 3 μl of the DpnI-treated PCR reaction with 5 μl sterile water and 2 μl In-Fusion enzyme mix. Incubate 15 minutes at 50°C.
5. When time to transform, an aliquot of frozen competent cells will be provided by the instructor. After thawing on ice, mix well by flicking the tube with your finger several times and transfer 50 μl to two 14 ml sterile culture tubes on ice.
6. Add 3 μl of your PCR reaction to one of the tubes and label it with a "+." Add nothing to the other tube and label it "-." Incubate on ice for 15 minutes.
7. Submerge the tube bottoms in a 42°C water bath for exactly 45 seconds and return to ice for two minutes.
8. Add 500 μl 2xYT media to each tube and incubate for 30 minutes at 37°C.
9. Mix the cell suspensions well, then transfer 100 μl from each tube to a LB-ampicillin plate and spread the cells thoroughly using the sterile glass beads by gently shaking the plates horizontally in all directions. Pour the beads off the plates into the collection bottle provided by the instructor.
10. Make sure the plates are labeled "+" and "-" (referring to the inclusion of the PCR product in the transformation) and with your group number or names. Place the plates in the 37°C incubator for overnight incubation.

Once the transformed bacteria grow into colonies (by the next day), there are some additional steps required before the mutants can actually be used. These steps are summarized below. Even though you will not be doing these in the lab due to time constraints, you are responsible for understanding them and their purpose.

- A couple of colonies will be used to start small liquid bacterial cultures.
- Plasmid DNA will be extracted from the bacterial cultures using a commercial DNA plasmid miniprep kit.

- The DNA will be submitted to a company for sequencing. You will get to analyze the sequencing data next week to determine if your mutagenesis reactions were successful in introducing the desired mutation.
- In the meantime, the teaching assistant will take successfully mutated plasmids and transform them into the appropriate bacterial cells for expressing the mutant protein so that they are ready for growth next week.

PART 2. MEDIA PREPARATION FOR NEXT WEEK

Bacterial growth medium must be prepared for next week when you will use *E. coli* to mass-produce your FoCdc14 protein. Each group should make 1 liter of 2xYT medium and split it into two 2-liter flasks.

Recipe for 2xYT Medium

Add the following to 1 liter of water and dissolve using a stir plate:

16 g Tryptone (milk protein digested with trypsin)

10 g yeast extract (digested yeast protein + soluble vitamins/molecules)

5 g NaCl

Distribute 500 ml into two 2-liter glass culture flasks and cover with aluminum foil.

Label the flask with your group number using lab tape and a marker.

Autoclave the flasks (this sterilizes the medium)—the teaching assistant or the instructor will take care of this.

PART 3. MICROPIPETTOR TESTING

You might think of micropipettors as simple devices that just suck up liquid in one place and squirt it out in another. However, they are actually highly accurate and precise analytical instruments that need to be handled carefully and rigorously maintained. What are the objectives of this exercise?

- Identify any poorly functioning micropipettors so they can be replaced or serviced prior to the enzymology labs where accurate pipetting will be critical.
- Emphasize some important principles and terms associated with analytical instruments, including accuracy, precision, error, variability, and calibration.
- Ensure that all students understand proper use and care of the micropipettors. Your group will be assigned a set of micropipettors for the semester and will be responsible for their proper use and care.

The principle for this exercise is that pure water has a density of ~1.00 g/cm^3, or ~1.00 g/ml (note that temperature and altitude affect the density slightly, but we will ignore those effects today). Therefore, if you measure 1 ml with a well-calibrated micropipettor, it should weigh exactly 1.00 g. The deviation from 1 g reflects the accuracy of that micropipettor. When the accuracy of a micropipettor (or many other analytical instruments) becomes less than acceptable, it must be recalibrated. You can't calibrate the micropipettors today, but you will check their accuracy and remove any that are problematic. Much of what you do this semester will depend on accurate pipetting of small volumes. The better you are at pipetting accurately, the better your data will be.

1. Each group should determine the accuracy of one set of micropipettors, including a P-1000, P-200, and P-20. Make three independent measurements for each micropipettor and record the masses of the indicated volumes of pure water

in the provided activity document. Use the analytical balances in the lab. Don't forget to tare the balance before each measurement.

2. Calculate the accuracy of each micropipettor as the % error.
3. Determine the precision of your measurements by calculating the standard deviation.
4. If the error of any of the micropipettors are off by more than 5%, notify the instructor. They may need to be serviced or replaced.

Notes:

- Please show your work for at least one set of measurement calculations.
- Make sure all group members contribute equally to measurements.
- Calculate the mean and standard deviation manually so that you understand the calculations. Do not use a calculator function.
- Then, calculate the accuracy of each micropipettor as average % error. Use the equations for standard deviation and % error provided above in the "General Concepts of Analytical Measurements" section.
- A data recording page has been uploaded to your Week 2 data folder. Fill in your data on that form and submit one completed sheet per group (e.g., upload an image of the completed sheet), being sure to include the names of all group members and the group number.

DATA ANALYSIS AND SUBMISSION

The only data file to submit today is the results of your micropipettor testing. The images of your transformation plates provided to you tomorrow will be additional data for the week.

3. MOLECULAR BIOLOGY AND RECOMBINANT PROTEIN EXPRESSION

BACKGROUND AND THEORY

In biochemistry, proteins are often characterized and studied in pure form, free from all other molecular contents of the cells they normally function in. The isolation of proteins of interest in pure form is therefore an important part of many biochemical research workflows, including in the drug discovery industry. Recombinant protein expression refers to the use of recombinant DNA vectors to produce a protein of interest in a foreign host organism, usually to facilitate its rapid, economical purification in large quantities. Many proteins of biological or biomedical interest are produced at very low levels in their native organism. This makes purification from the natural source technically challenging or even impossible. If protein variants, such as site-directed mutants, need to be engineered and purified to characterize effects of mutations on protein function, then recombinant expression systems are often the only practical option.

RECOMBINANT DNA AND MOLECULAR CLONING

What is a recombinant DNA vector? The term "recombinant DNA" refers to a DNA molecule built from DNA sequences derived from multiple sources and combined into a single engineered construct for research and biotechnology applications. The term "vector" refers to a vehicle used to introduce foreign DNA into a host cell. In many cases the vector is a plasmid, a circular DNA molecule, discussed more below. But it can also be a virus or other type of DNA molecule. With modern molecular cloning technologies, DNA vectors can be engineered to express proteins of interest at very high levels in organisms that are fast-growing, inexpensive, and easy to culture at scale. This allows large amounts of pure protein to be generated much more easily for biochemical analysis. Proteins made by expressing foreign genes from a recombinant DNA vector inserted into another organism are referred to as "recombinant proteins."

Bacteria, especially *Escherichia coli*, are the most common hosts used for recombinant protein production because they grow extremely quickly and because many recombinant DNA reagents, systems, and optimized expression tools have already been developed for them. Eukaryotic systems are also used, including budding yeast (*Saccharomyces cerevisiae*), methylotrophic yeast (*Pichia pastoris*), and insect cells such as Sf9 infected with engineered baculovirus expression vectors. Eukaryotic hosts, such as yeast and insect cells, are preferred in cases where the target protein requires eukaryote-specific chaperones or posttranslational modifications for proper folding or function, or if the protein is very large or otherwise difficult to express in bacteria.

The generation of recombinant DNA vectors is performed through molecular cloning techniques. One widely used classical method relies on restriction endonucleases that cut DNA at specific recognition sequences, coupled with DNA ligase, which joins complementary DNA ends to produce chimeric molecules. Other cloning systems now exist that do not require restriction enzymes. The Gateway system (marketed by Invitrogen) uses recombination enzymes to swap DNA segments between DNA molecules,[1] while Gibson Assembly uses annealing of homologous regions between insert DNA and vector DNA to assemble multiple fragments simultaneously into one plasmid.[2] The In-Fusion system (Takara Bio) is a more recent and simplified variation of Gibson Assembly.[3] The design of cloning systems such as In-Fusion and Gibson Assembly that do not rely on restriction enzymes or specific enzyme recognition sites was a major advance, as it allowed (1) any foreign DNA sequence of interest to be inserted anywhere in a vector molecule and (2) the assembly of complex recombinant molecules containing multiple inserts in a single step. This week's lecture has additional information on general features of different molecular cloning technologies.

The vectors used for recombinant protein expression in *E. coli* are plasmids. Plasmids are naturally occurring, circular, double-stranded extrachromosomal DNA molecules that exist in many bacteria, yeast, and other organisms and frequently encode genes conferring antibiotic resistance. Scientists have modified naturally occurring plasmids so they can be easily used to express foreign genes in *E. coli* and other host species. For expression of recombinant proteins in *E. coli*, plasmid expression vectors must contain (1) an origin of replication enabling propagation in actively dividing cells; (2) a selectable marker, typically an antibiotic resistance gene, allowing selection of plasmid-containing cells among a background of plasmid-free cells; and (3) a regulatable promoter capable of selectively driving high-level transcription of just the target gene of interest (Figure 3.1). For classical cloning purposes, plasmids also needed a molecular cloning site for insertion of the foreign gene to be expressed. These are not necessary for In-Fusion, Gibson Assembly, and related cloning systems. Many recombinant expression plasmids also contain the coding sequence for one or more affinity tags that are translated along with the target protein, enabling facile purification, as discussed in detail in Chapter 4. Plasmid sequences can be viewed and manipulated in a variety of software programs, allowing researchers to custom design recombinant molecules.

RECOVERY AND CONFIRMATION OF RECOMBINANT PLASMIDS

In cloning workflows (Figure 3.2), the final assembled recombinant plasmid molecules are recovered in pure form by transformation into bacterial cells, which are then plated on agar media containing a selective antibiotic drug. The term "transformation" describes the forced, stable insertion of a plasmid into a host cell. The inclusion of an antibiotic ensures that only the rare cells successfully transformed with a plasmid (containing a gene for resistance to the antibiotic) grow. Growth of nontransformed cells is prevented by the antibiotic. Colonies from the transformation plate, consisting of clonal cells that each contain the identical plasmid, are used to start small liquid bacterial cultures from which the plasmid can be purified. Analytical procedures are needed to confirm that the correct recombinant molecule was successfully made. These procedures include analytical restriction mapping, polymerase chain reaction, and DNA sequencing.

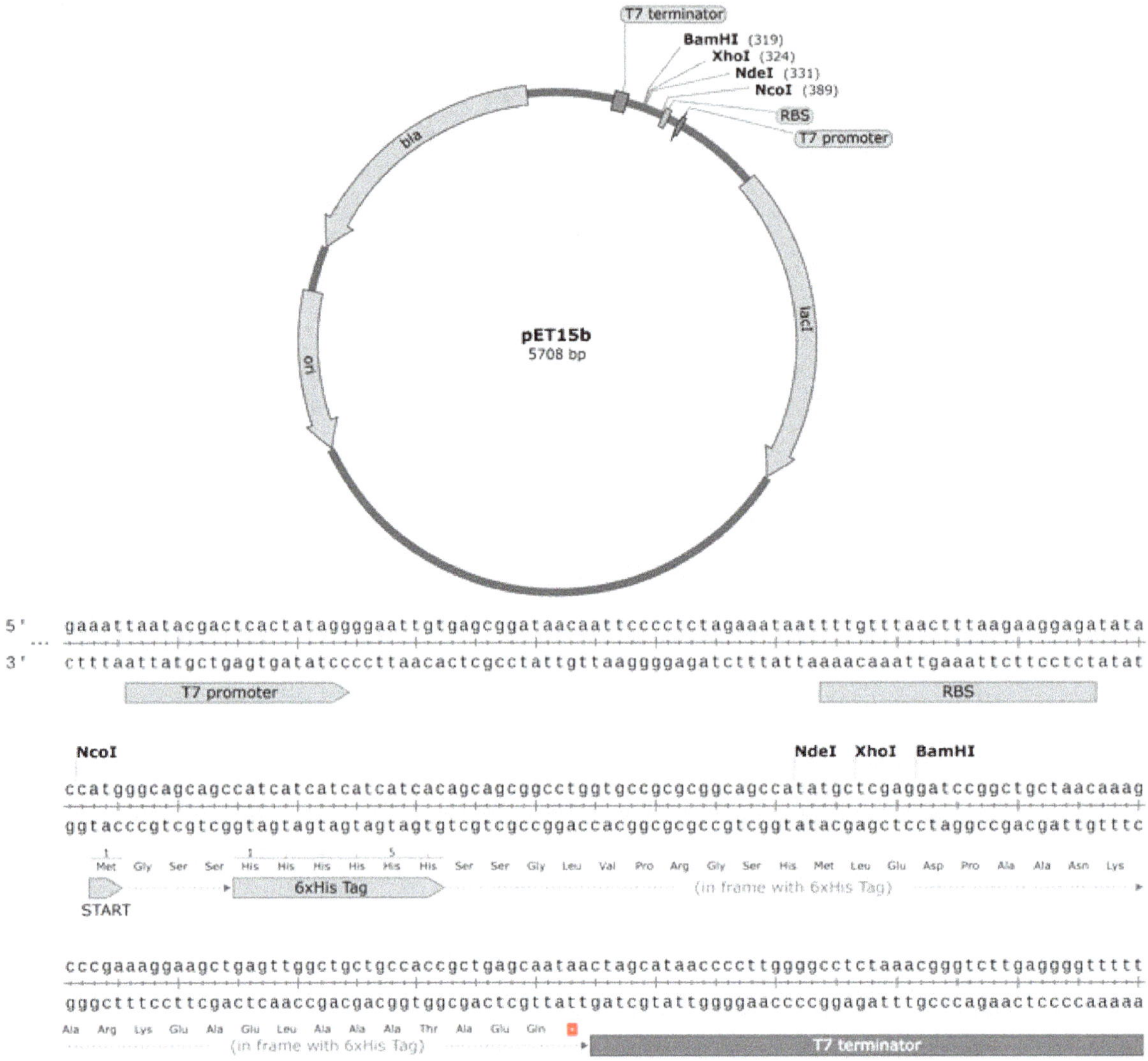

FIGURE 3.1. Plasmids for recombinant protein expression, such as pET15b, must have an origin of replication for maintenance in bacterial cells (ori), an antibiotic resistance gene such as *bla* (encoding β-lactamase, conferring ampicillin resistance), and a regulatable promoter for inducible target gene transcription (T7 promoter with associated transcriptional terminator). In pET15b the product of the *lacI* gene prevents leaky expression in the absence of induction. The sequence at bottom shows the cloning and expression region in more detail. RBS is the ribosome binding site, where translation of the mRNA will initiate. Restriction enzyme sites (NcoI, NdeI, XhoI, BamHI) are available for insertion of genes to be expressed by conventional molecular cloning, allowing fusion with a 6xHis affinity tag.

THE T7 EXPRESSION SYSTEM

The inducible T7 RNA polymerase system is the most widely used recombinant protein expression platform in *E. coli.*[4] In typical pET vectors (such as those originally developed by Novagen), foreign genes are cloned into a defined region flanked by T7 promoter and terminator sequences. The host strain encodes the T7 RNA polymerase gene under the control of the lac operon's *lac* promoter or, alternatively, the araBAD operon's *pBAD* promoter (Figure 3.3).[5] Induction using the lactose analog, IPTG (lac system), or L-arabinose (*pBAD* system)

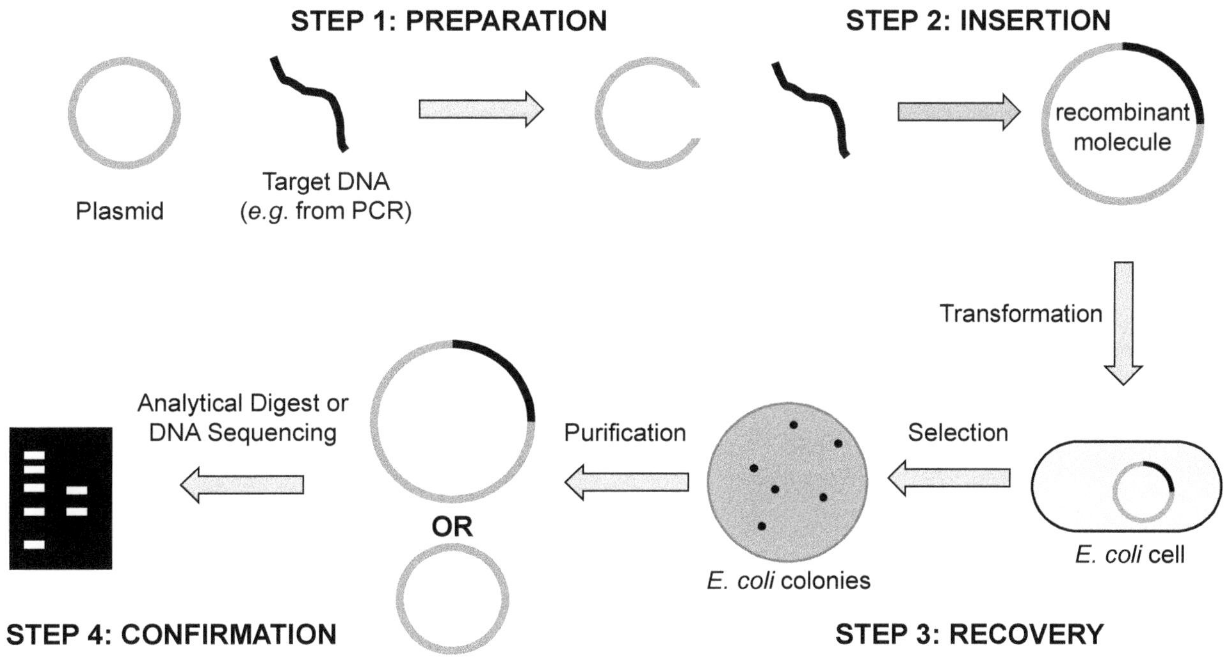

FIGURE 3.2. A standard molecular cloning workflow begins with generation of the desired target DNA, usually by synthesis or PCR amplification. The second step involves inserting the target DNA fragment into the desired location within the plasmid. A variety of methods for generating this recombinant plasmid are available. The recombinant plasmid must then be recovered in pure form, separated from the original plasmid. This is accomplished by transforming the reaction mixture from step 2 into *E. coli* cells, using the plasmid's antibiotic resistance marker for positive selection on agar plates containing the antibiotic. Individual colonies are picked and amplified in liquid culture, and their plasmid DNA is extracted and purified. Analytical procedures such as DNA sequencing and agarose gel-based restriction enzyme analysis are used to determine if the correct recombinant molecule has been recovered.

activates expression of the bacteriophage T7 RNA polymerase, which then selectively drives high-level transcription of the foreign target gene from the T7 promoter in the pET plasmid, enabling mass production of the recombinant protein (Figure 3.4). Your recombinant protein will be expressed using the pET15b vector shown in Figure 3.1 and the *pBAD* version of the T7 expression system.

Induction of protein expression is best done in actively growing bacterial cultures. Growth of cells in culture follows an exponential pattern (Figure 3.5), with an early lag phase of slow growth, a later log phase of rapid growth, and finally a stationary phase when nutrients are depleted or metabolic by-products inhibit growth.

Protein expression works best during the rapid log phase growth. To ensure that cultures are in this rapid growth phase when target protein expression is induced with L-arabinose, a method to monitor culture growth is needed. This is conveniently accomplished by measuring the culture's ability to absorb light using a spectrophotometer. Typically, for microbial growth measurements a light wavelength of 600 nm is used. Absorbance of light at 600 nm increases as the cell density increases (see Figure 3.5), with optimal log phase growth corresponding to an absorbance value of ~0.5–1. You will use a spectrophotometer today to measure culture absorbance to make sure your cells are in log phase growth when you induce expression and also to normalize the cell density of your uninduced and induced culture samples for comparative analysis by protein electrophoresis in Week 5. You will learn more about spectrophotometry as a versatile analytical tool in a lecture later in the course.

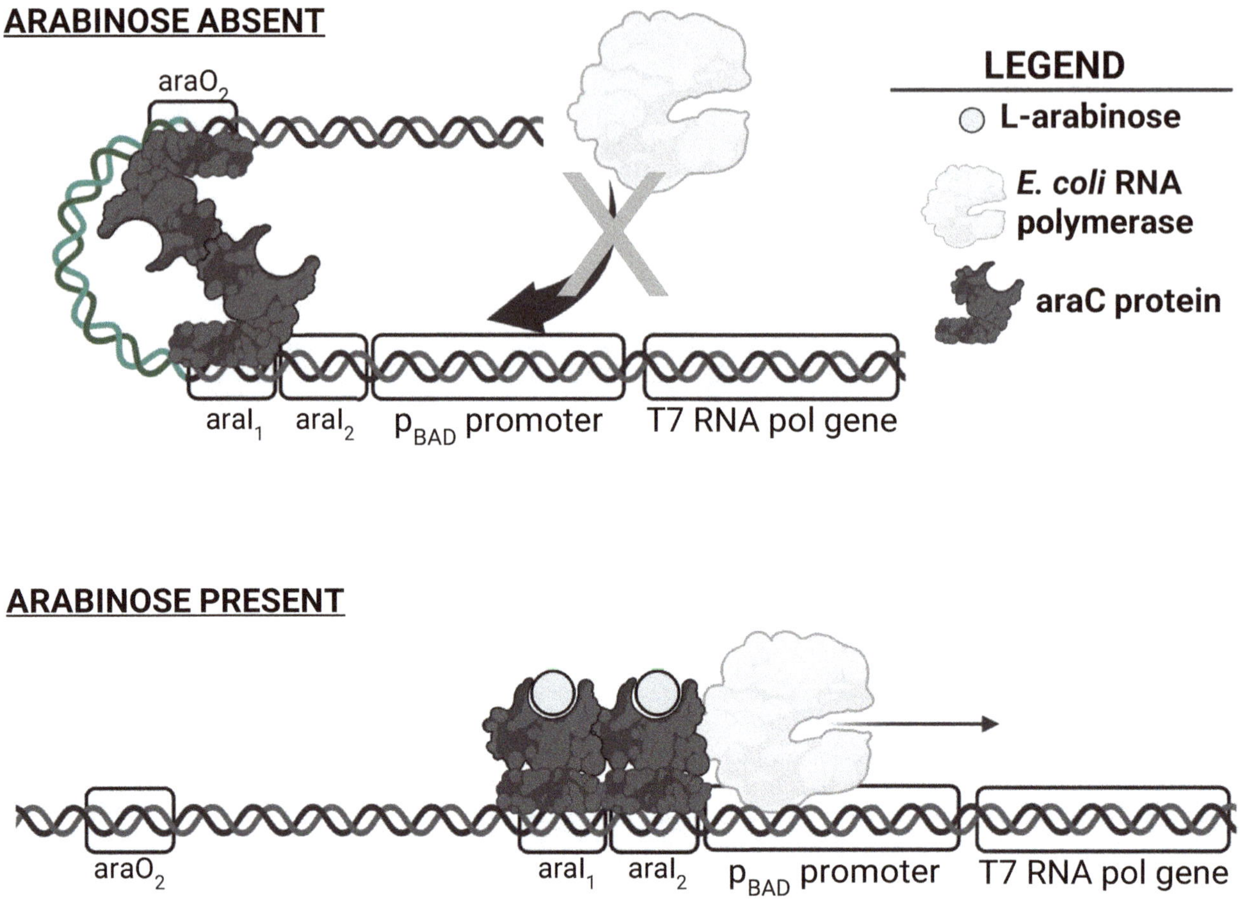

FIGURE 3.3. In the araBAD operon, the p_{BAD} promoter is controlled by the AraC protein. In the absence of L-arabinose (top), the AraC dimer adopts a conformation that promotes binding to two distal DNA regulatory sites, $araO_2$ and $araI_1$, causing a hairpin structure in the DNA that physically blocks access by *E. coli* RNA polymerase. When L-arabinose binds to AraC (bottom), the dimer changes conformation, promoting binding to the adjacent $araI_1$ and $araI_2$ sites, opening the DNA hairpin, and allowing recruitment and binding of RNA polymerase to p_{BAD}. In the engineered T7 expression system this results in transcription of the T7 RNA polymerase gene. Created in BioRender. Hall, M. (2026) https://BioRender.com/5fvbzmo.

OBJECTIVES

The primary goal this week is to grow a large culture of *E. coli* cells containing the plasmid for expressing recombinant *F. oxysporum* Cdc14 (FoCdc14) and induce expression by addition of L-arabinose to the culture. The induction will proceed overnight at 25°C, and the cells with expressed FoCdc14 will be harvested by centrifugation the next morning by the instructor and the teaching assistant and stored as a cell pellet at -80°C until next week. In addition, small cell samples before and after induction will be saved for analysis of the induction by protein electrophoresis. Note that one or two groups will be assigned to express and purify the predicted catalytic mutant version of FoCdc14 that was the subject of the mutagenesis efforts last week. Preparations will be shared so that each group has wild-type and mutant FoCdc14 for enzymatic comparison in a later lab.

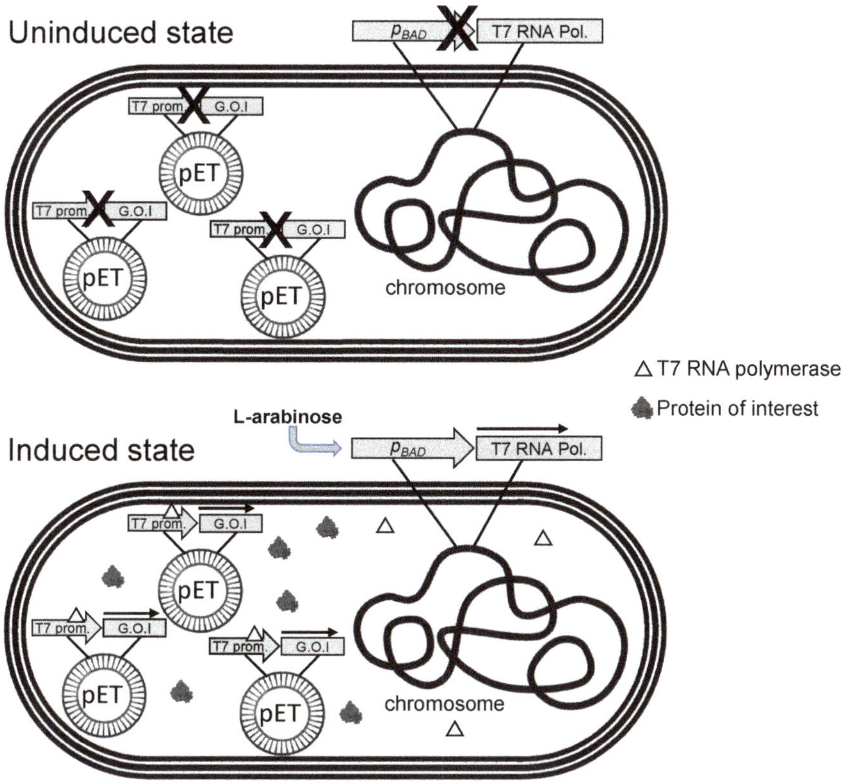

FIGURE 3.4. The T7 expression system ensures that the gene of interest (GOI) is not transcribed and its encoded protein is not produced until the inducing agent is added. In the absence of L-arabinose in the araBAD system (top), expression of T7 RNA polymerase from p_{BAD} (an engineered gene cassette inserted into the bacterial chromosome) is repressed, as shown in Figure 3.3. As a result, the GOI inserted downstream of a T7 RNA polymerase-specific promoter (T7 prom) in the pET plasmid is also not expressed. Upon addition of L-arabinose (bottom), p_{BAD} is activated, leading to production of T7 RNA polymerase, which then binds its promoter on the pET plasmid and transcribes the GOI, resulting in production of the recombinant protein of interest. Created, in part, in BioRender. Hall, M. (2026) https://BioRender.com/47nur2v.

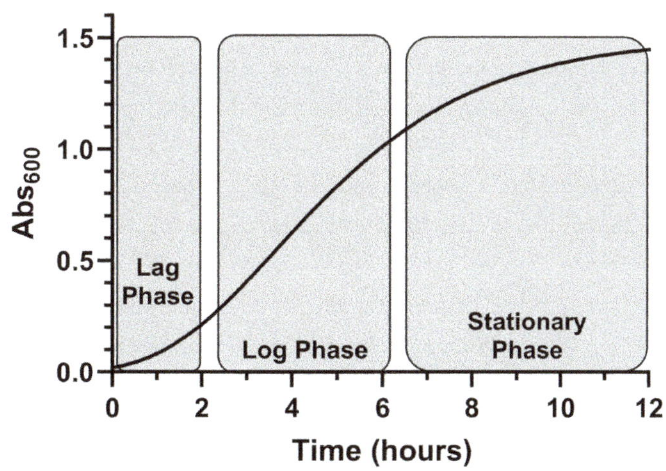

FIGURE 3.5. An example plot of absorbance at 600 nm (Abs_{600}) versus time for a liquid bacterial culture. After an initial lag phase of relatively slow growth, the bacterial cells divide exponentially, a period called log phase growth. When nutrients are depleted and toxic metabolic products accumulate, growth gradually comes to a stop in the stationary phase.

There are two additional secondary goals for this week. First, several solutions needed for the protein purification, electrophoresis, and immunoblotting labs the next few weeks will be prepared. This provides practice making correct calculations for achieving desired concentrations of solution ingredients. It is critical that these solutions are made correctly; otherwise, procedures won't work. Second, you will analyze DNA sequencing data obtained from plasmid miniprep samples generated from your site-directed mutagenesis efforts last week. You will determine whether your polymerase chain reaction to create the putative inactivating mutation in the FoCdc14 catalytic cysteine within the pET15b expression plasmid was successful or not.

LAB INSTRUCTIONS

PART 1. BACTERIAL GROWTH AND PROTEIN EXPRESSION

1. In the morning lab period, start large cultures growing.
   A. To each large media flask (0.5 L media volume), add ampicillin to a final concentration of 100 µg/ml. The ampicillin stock is 100 mg/ml. Make sure you use proper sterile technique! You are growing cells that can easily become contaminated. The instructor will review sterile technique. Also, make sure you know why you are adding ampicillin to the growth media.
   B. Add 5 ml of saturated *E. coli* cell culture provided to your group to each of the large media flasks. You have now inoculated (started) your cultures for protein expression. These cells contain the pET15b plasmid for expressing FoCdc14 using the L-arabinose-inducible T7 RNA polymerase system.
   C. Mix the flasks well and then remove a 1 ml sample from one of them and transfer it to a plastic cuvette provided to your group. Transfer 1 ml of sterile 2xYT media to another cuvette to use as a blank for the spectrophotometer.
   D. Measure and record the starting absorbance of your culture at 600 nm using the available spectrophotometers. The instructor or the teaching assistant will demonstrate how to use these.
   E. When ready, place your flasks in one of the shaking incubators. Make sure your flasks are labeled with your group number. The cultures will be grown until late in the afternoon lab period, when you will induce overexpression of your target protein.
2. During the afternoon lab period, initiate protein overexpression and save an uninduced cell sample.
   A. Remove a 1 ml sample from one of your large cultures and transfer it to a plastic cuvette provided to your group. Transfer 1 ml of sterile 2xYT media to another cuvette to use as a blank for the spectrophotometer.
   B. Measure the absorbance of your culture again at 600 nm using the spectrophotometers. Record the absorbance value.
   C. Continue checking absorbance at 30 min intervals until the Abs_{600} is near 0.75. This represents log phase growth. Check the absorbance of the cultures in both flasks now to make sure they are roughly the same density and are ready to induce.
   D. Once the Abs_{600} is relatively close to 0.75 (it doesn't need to be exact—between 0.6 and 1 is okay), remove another 1 ml sample from one of your large cultures and transfer it to a microfuge tube. Label the tube "Preinduction" with your group number on it.
   E. After removing the 1 ml preinduction sample, add L-arabinose from the 20% stock provided to you to both large culture flasks at a final concentration of 0.1% and place them back in the shaking incubator. This will initiate overexpression of your target protein from the plasmid DNA. The cells will be harvested by the instructor or teaching assistant early tomorrow morning and kept frozen until next week.

F. Spin the 1 ml preinduction sample down in a microcentrifuge for 2 minutes at 7,000 rpm (partner with another group so the centrifuge is balanced). Discard the supernatant into the liquid waste container using a pipettor.
G. Resuspend the preinduction cell pellet in SDS-PAGE sample dye (provided to each group) according to the following formula: 200 μl x Abs_{600} reading.
H. Heat the sample in a heat block at 95°C for 5 minutes.
I. Provide the preinduction cell extract sample to your instructor or teaching assistant to store at −20°C for analysis by electrophoresis later. We will be comparing the preinduction and postinduction extracts by SDS-PAGE in two weeks. You should expect to see a strong protein band at the expected size of your target protein appear in the postinduction sample, but the band should be absent from the preinduction sample. This will confirm that the expression system worked as intended and that your protein has been made by the bacterial cells.

3. After protein overexpression is complete (this will be completed by the teaching assistant or the instructor the following morning):
   A. Remove a 1 ml sample from the large flask and transfer it to a microfuge tube. Label the tube "Postinduction" with your group number on it.
   B. Remove a 0.5 ml sample and transfer it to a plastic cuvette containing 1.5 ml 2xYT medium.
   C. Measure the absorbance at 600 nm of the induced culture, multiply by the dilution factor of 4, and record the value. The dilution is necessary to get an accurate absorbance when the cell culture becomes this dense.
   D. Prepare the postinduction cells in SDS-PAGE sample dye exactly as described for the preinduction cells.

PART 2. DNA SEQUENCING DATA ANALYSIS

Remember that after last week, the plasmid DNA was recovered from a couple of your transformation colonies using the DNA miniprep procedure and shipped to a company for DNA sequence analysis. Your job is to analyze the data that came back and determine if the desired mutation is present in the plasmids that were recovered from the transformed *E. coli* cells. Just because there are colonies on a transformation plate does not automatically mean the plasmid in those cells is the correct, intended one. This must be confirmed.

1. Open the SnapGene Viewer program on your computers. Then, open the "pET15b-FoCDC14 plasmid sequence" file that has been uploaded to your Week 3 data storage folder ahead of time. This file contains the exact sequence of the expression plasmid that you are using and that was the template for your mutagenesis reactions last week. Note that the FoCDC14 coding sequence in this plasmid is different than the natural coding sequence you worked with in Week 1 because it has been codon-optimized for expression in *E. coli* and has been truncated some to remove unnecessary sequence at the 3′ end.
2. Select the entire *FoCDC14* coding sequence by clicking on it in Sequence view to highlight it and then selecting Edit → "Copy Top Strand Bases."
3. Access Clustal Omega (the site that you used to generate the multisequence alignment of Cdc14 orthologs in Week 1) at https://www.uniprot.org/align/. Paste in the coding sequence of your gene from Week 1, preceded by a line with the text ">FoCDC14_CDS." Hit return and then enter another text line ">Mut_Seq_Data," hit return again, and paste in the text from one of the sequencing analysis files provided in your group's Week 3 data storage folder. Hit "Run Align" at the bottom.
4. To evaluate the resulting alignment, you will need to find the catalytic site that you were trying to alter, using the SnapGene file and your previous alignment of Cdc14 orthologs where you identified the active site. You will need to

reference this to determine if you successfully achieved the desired sequence substitution at this site to convert the catalytic cysteine codon to a serine codon.

5. Perform this alignment for each of your data files. Using the SnapGene file for reference, assess whether or not you successfully created the intended mutation in the expression plasmid. Create and save a Word file that contains a copy of each sequence alignment, with the mutation site highlighted and your conclusion about the success of the mutagenesis stated clearly. You will need to submit this file in the course learning management system.

PART 3. MAKING SOLUTIONS

Concentration is defined as an amount per unit of volume. It can be defined in different ways using different units. You need to understand the difference between them and also how to calculate reagent amounts to make accurate concentrations in your solutions. You should already know how to do this, so hopefully this is nothing more than a review (but it is always good to get more practice!). **Concentrations and units are going to be very important in this course, as they always are in biochemistry.**

1. MOLAR CONCENTRATION

Molarity is the number of moles of solute per liter of solution. To make solutions with molar concentrations, you need to know the formula weight of the solute. For example, NaCl has a formula weight of 58.45 grams/mole. Therefore, a 1 M solution of NaCl would contain 58.45 grams of NaCl per liter. A simple way to calculate the mass of a chemical needed to obtain a given molarity is

Mol Wt (g/mole) x desired M (moles/L) x desired volume (L) = g chemical

2. PERCENT SOLUTION

There are multiple ways to make a percent solution. One is percent by weight (% wt/wt). For example, a 10% solution would contain 10 g of solute per 100 g of solution (10g/100g x100 = 10%). The second and more common is percent by volume (% wt/vol). For example, a 10% solution would contain 10 g solute per 100 ml of solution. In some cases you might make a vol/vol solution if your chemical is in liquid form. For example, a 10% acetic acid solution would be 10 ml of glacial acetic acid added to 90 ml water (10 ml per 100 total ml of solution).

3. WEIGHT PER VOLUME

A common way to express concentration is simply the mass of a chemical per unit volume. For example, if you require a 10 mg/ml solution of a chemical, you need to add 10 mg of chemical for each ml of solution being made. To find the mass of chemical needed, simply multiply the final concentration (10 mg/ml) by the desired final volume.

You are going to make several solutions to be used in later labs that represent the different units of concentration. Each group should make the calculations and check them with the professor or the teaching assistant. These calculations must be submitted as a file in the course learning management system as part of the data submission assignment for this week. Note that a separate Word file containing all of the solution recipes, shown below in the section "Solution Recipes and Calculations Form," has been provided for filling in and submitting your group's calculations. Once all groups are done with the calculations, assignments for making the solutions will be distributed. Each group will make one or two solutions. Remember that the final pH for many of the solutions must be verified. Don't assume that the pH is correct without checking it.

DATA ANALYSIS AND SUBMISSION

Before leaving today, make sure that your group submits the following two files to the course learning management system:

1. The document with complete solution calculations
2. Results and conclusions from analysis of your DNA sequencing data

In addition, make sure you have properly labeled your preinduction *E. coli* extract sample and given it to the instructor or the teaching assistant for storage at −20°C. Finally, ensure that you have finished making, labeling, and storing your assigned solutions.

SOLUTION RECIPES AND CALCULATIONS FORM

SDS-PAGE Running Buffer—weight per volume. Make 1 L per two groups

3 mg/ml Tris base

14.4 mg/ml glycine

1 mg/ml sodium dodecyl sulfate (SDS)

If made correctly, pH should be between 8.3 and 8.5. Check to make sure.

Calculations

Tris base (g) ________________________

Glycine (g) ________________________

SDS (g) ________________________

Final pH = ________________________

PBS (phosphate-buffered saline)—Molarity. Make 1 L per three groups

137 mM NaCl

2.7 mM KCl

4.3 mM Na_2HPO_4

1.47 mM KH_2PO_4

Adjust pH to 7.5 if necessary

Calculations

NaCl (g) ____________________

KCl (g) ____________________

Na_2HPO_4 (g) ____________________

KH_2PO_4 (g) ____________________

Final pH = ____________________

Lysis Buffer for Nickel Affinity Purification—mix. Make 0.5 L

25 mM Tris-HCl pH 7.5 (dilute from a 0.5 M stock solution at pH 7.5)

500 mM NaCl

10 mM imidazole

10% glycerol

0.1% Triton X-100 (dilute from 10% stock solution)

Calculations

Tris-HCl (ml of 0.5 M stock) ____________________

NaCl (g) ____________________

10% Triton X-100 solution (ml) ____________________

Glycerol (ml) ____________________

Imidazole (g) ____________________

Final pH = ____________________

Nickel Buffer A—mix. Make 1 L per two groups

25 mM Tris-HCl pH 7.5

500 mM NaCl

10% glycerol

Calculations

Tris-HCl (ml of 0.5 M stock) ____________________

NaCl (g) ____________________

Glycerol (ml) ____________________

Final pH = ____________________

Nickel Buffer B—mix. Make 1 L per three groups

25 mM Tris-HCl pH 7.5

500 mM NaCl

10% glycerol

250 mM imidazole

Calculations

Tris-HCl (ml of 0.5 M stock) ____________________

NaCl (g) ____________________

Glycerol (ml) ____________________

Imidazole (g) ____________________

Final pH = ____________________

Enzyme Storage Buffer—mix. Make 2 L

25 mM Tris-HCl pH 7.5

300 mM NaCl

0.1% 2-mercaptoethanol (stock is 100%)

2 mM EDTA (dilute from 500 mM stock)

40% glycerol

Calculations

0.5 M Tris-HCl stock (ml) ____________________

NaCl (g) ____________________

2-mercaptoethanol (ml) ____________________

500 mM EDTA (ml) ____________________

Glycerol (ml) ____________________

4. PROTEIN PURIFICATION

BACKGROUND AND THEORY

CHROMATOGRAPHY FOR PROTEIN PURIFICATION

Chromatography is the standard methodology for purifying individual proteins to enable rigorous biochemical and biophysical characterizations in the absence of other interfering biomolecules. It provides separation of proteins based on differences in their physical and/or chemical properties, such as charge and size. Chromatography involves passing a mobile phase containing soluble proteins through a stationary phase, which provides the separating power via differential retention. The standard chromatography format is the column (Figure 4.1). Stationary phase particles are packed inside the column, and an inlet and outlet allows flow of the mobile phase through the column for interaction with the stationary phase. A pump system is often used to push the mobile phase through the stationary phase, and most chromatography systems contain some type

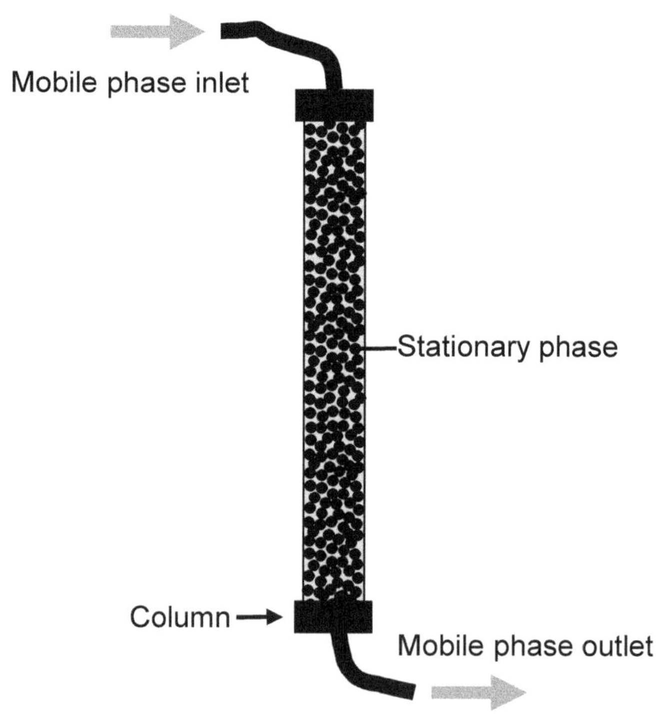

FIGURE 4.1. A chromatography column for protein purification contains stationary phase particles with chemical and/or physical properties that mediate the differential retention of proteins. The liquid mobile phase containing analytes (proteins) to be separated passes through the column from one end to the other allowing interaction of analytes and stationary phase.

of detector to monitor proteins eluting from the column and a device to collect the separated protein fractions (Figure 4.2).

The stationary phase is usually particles made from an inert polymer to which functional groups are attached that interact with the proteins in the mobile phase. For example, in ion exchange chromatography the functional groups are either negative or positive chemical groups on the particles that interact predominantly with amino acid side chains of opposite charge on the mobile phase proteins. For cation exchange columns, proteins with strong positive charge will bind the stationary phase most tightly and therefore take the longest to elute from the column (Figure 4.3).

In contrast, the functional group for hydrophobic interaction chromatography is usually a nonpolar hydrocarbon, and the slowest eluting proteins will be those with the most nonpolar character. Elution of proteins from stationary phases such as ion exchange and hydrophobic interaction is often facilitated by gradually changing the ionic conditions of the mobile phase. For ion exchange, increasing the salt concentration over time sequentially elutes proteins in order of binding strength as the salt ions compete for binding to the stationary phase. For hydrophobic interaction, decreasing the salt concentration weakens the strength of nonpolar interactions with the stationary phase. Size exclusion chromatography is different in that the stationary phase particles don't have functional groups that bind proteins but instead have pores and channels that exclude larger proteins but allow passage of smaller proteins. Size exclusion chromatography will be covered in more detail in Chapter 8, where it will be used as an analytical method to characterize protein oligomeric state.

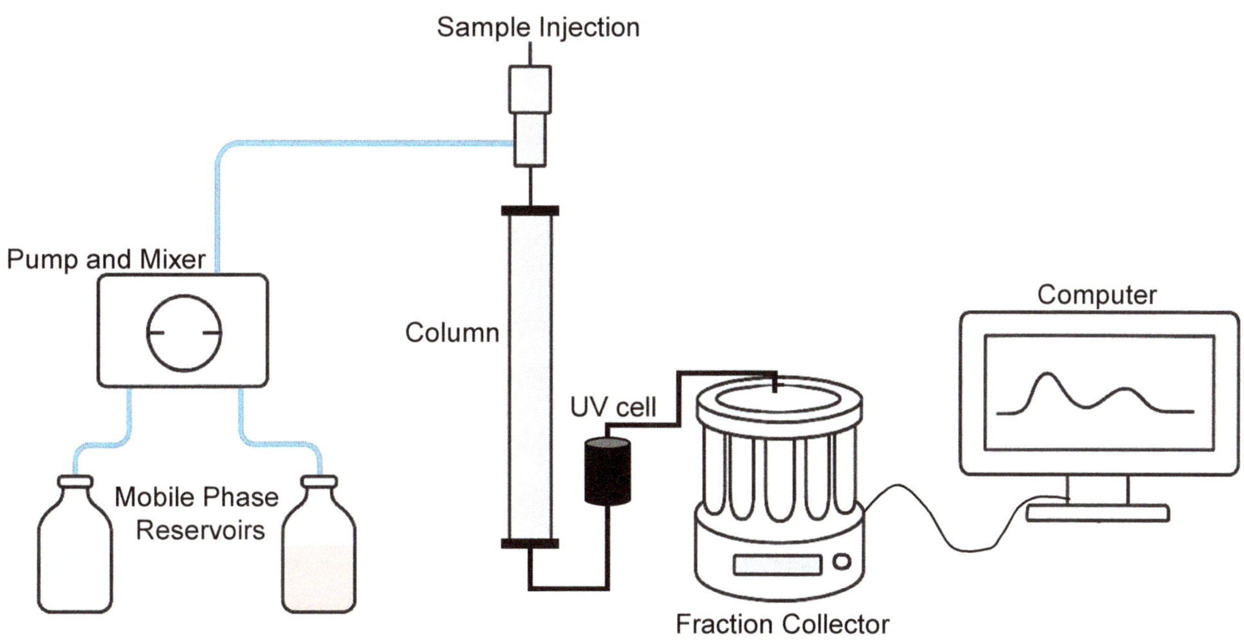

FIGURE 4.2. Diagram and flowpath for a typical chromatography system used for protein purification. A pump powers flow through the system, and a mixing valve combines liquid from multiple mobile phase reservoirs at user-defined ratios. A sample injection port allows introduction of an analyte mixture to the column. Flow passes through the column and a UV cell for protein detection by absorbance spectroscopy, ending at a fraction collector, which isolates discrete aliquots of the column eluate. The entire system is controlled by a computer.

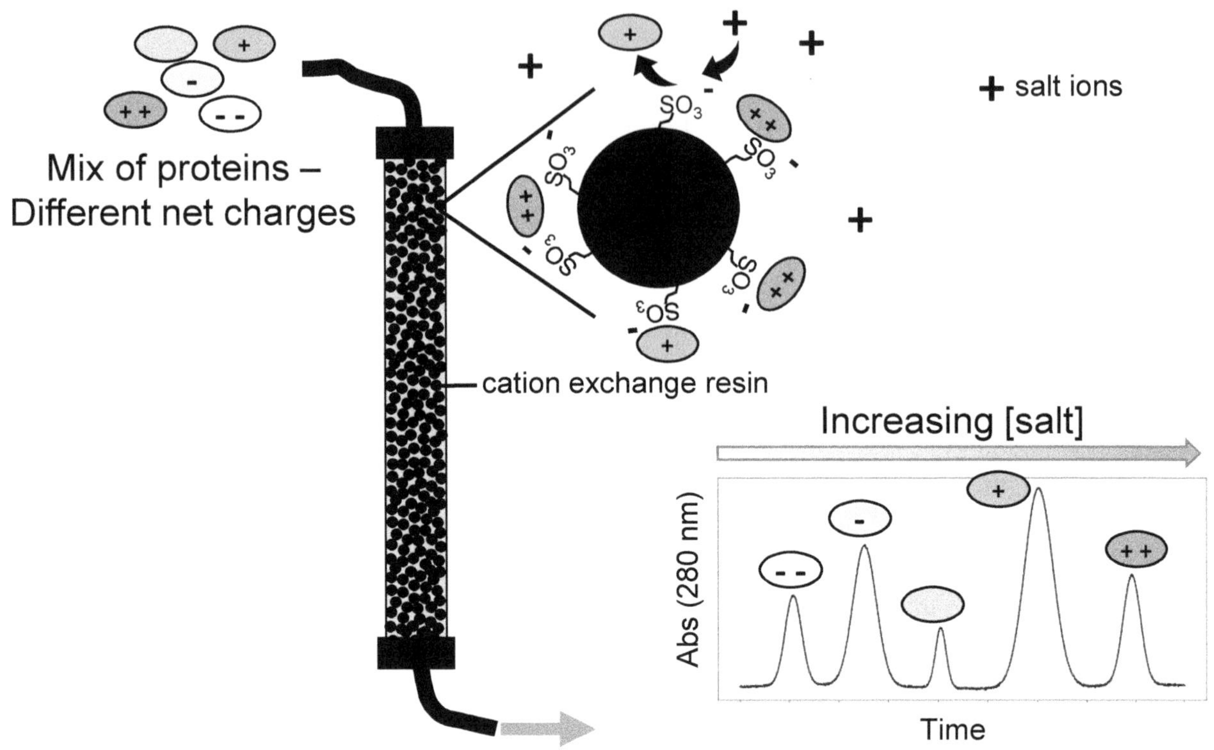

FIGURE 4.3. In cation exchange chromatography, the stationary phase particles bear negatively charged functional groups (sulfonate in this case) that bind positively charged regions on mobile phase proteins. Strength of binding to the stationary phase depends on overall protein charge state, with the most negative proteins having little to no interaction. Bound proteins can be competitively eluted in order of increasing positive charge and strength of binding by gradually raising the salt concentration in the mobile phase. Proteins with greatest net positive charge will require the highest salt concentration for displacement.

PURIFICATION OF PROTEINS BY AFFINITY CHROMATOGRAPHY

Affinity chromatography is a powerful purification strategy that takes advantage of high affinity binding between a protein and a ligand (Figure 4.4). In this method, the ligand is covalently bound to the column matrix, typically agarose beads, as the stationary phase. A sample containing the target protein, often an extract from a biological source, is passed through the column, allowing the target proteins to interact with and bind tightly to the ligand molecules. Other unwanted molecules are washed away, and then the pure target molecules are recovered by disrupting their interaction with the immobilized ligand. This is often done by competition with free ligand in the mobile phase.

Molecular cloning and recombinant DNA technology allow the easy creation of gene fusions in which a gene encoding a protein of interest is fused to DNA sequence encoding an affinity tag, resulting in the production of a single polypeptide consisting of both the protein of interest and the tag. Because they are linked, purification of the affinity tag by binding to an immobilized ligand results in purification of the protein of interest. These fusion proteins are often expressed in heterologous organisms (e.g., human genes can be cloned into bacterial expression plasmids to overproduce and purify the proteins from bacterial cultures). You received an overview of molecular cloning techniques last week.

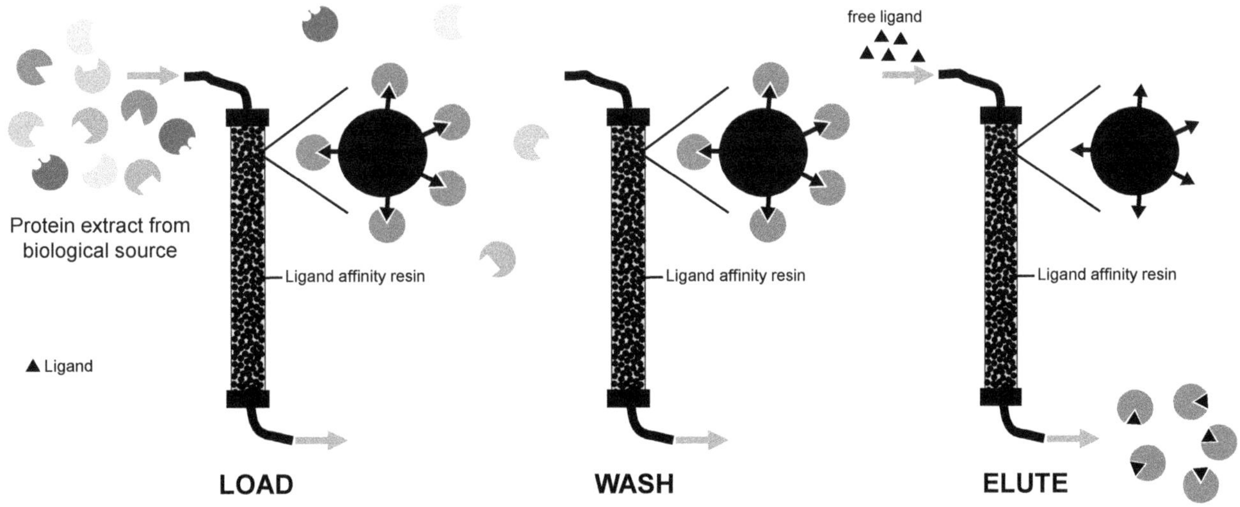

FIGURE 4.4. In affinity chromatography, a ligand is covalently linked to the stationary phase, allowing highly selective retention of proteins that recognize and bind the ligand. After washing away unbound proteins, the target ligand-binding protein is recovered by competitive elution with free ligand in the mobile phase.

One common strategy for affinity purification is to fuse the gene of interest to the sequence for a eukaryotic glutathione S-transferase (GST), to take advantage of the high affinity it has for its substrate, glutathione (GSH). GSH is a tripeptide of sequence glutamate-cysteine-glycine. It is the ligand for affinity chromatography purposes and is covalently attached to agarose beads as the stationary phase. The natural biological function of GST enzymes is detoxification of aromatic compounds by modifying them with GSH. There are many other examples of similar affinity-based purification systems in use.

A unique but common strategy for affinity purification is to fuse the gene of interest to six consecutive histidine codons. The six histidines of the 6His tag bind very tightly to divalent metal ions, such as Ni^{2+} or Co^{2+}, that are chelated (immobilized) on a solid phase support, commonly via the functional group nitriloacetic acid (Figure 4.5). Although this isn't a natural protein-ligand interaction, the concept is the same, with Ni^{2+} acting as the ligand for the 6His-tagged protein. The expression plasmid you are using, pET15b, encodes a 6His tag for fusion to the amino terminus of target proteins like your FoCdc14. You will use Ni^{2}-agarose columns today to purify the 6His-FoCdc14 fusion protein.

To purify the target protein, each group will make a soluble protein extract from the *E. coli* cells grown last week that have overproduced your 6His-tagged FoCdc14 protein. The extract will be passed through the Ni^{2+}-agarose affinity column to allow binding of the fusion protein to the immobilized Ni^{2+} ions. Other proteins will be washed out, and your target 6His fusion protein will be eluted (hopefully) with a high concentration of imidazole. Imidazole is the side chain on histidine amino acids and competes with your 6His protein for binding to the Ni^{+2}, thereby releasing your target protein back into solution, which is then collected.

At the end of the day you will transfer your protein to a suitable storage buffer by dialysis. In subsequent weeks you will evaluate how well the purification worked by SDS-PAGE and immunoblotting. You will also learn how to calculate such things as concentration of your protein and the purity and yield obtained from your purification procedure.

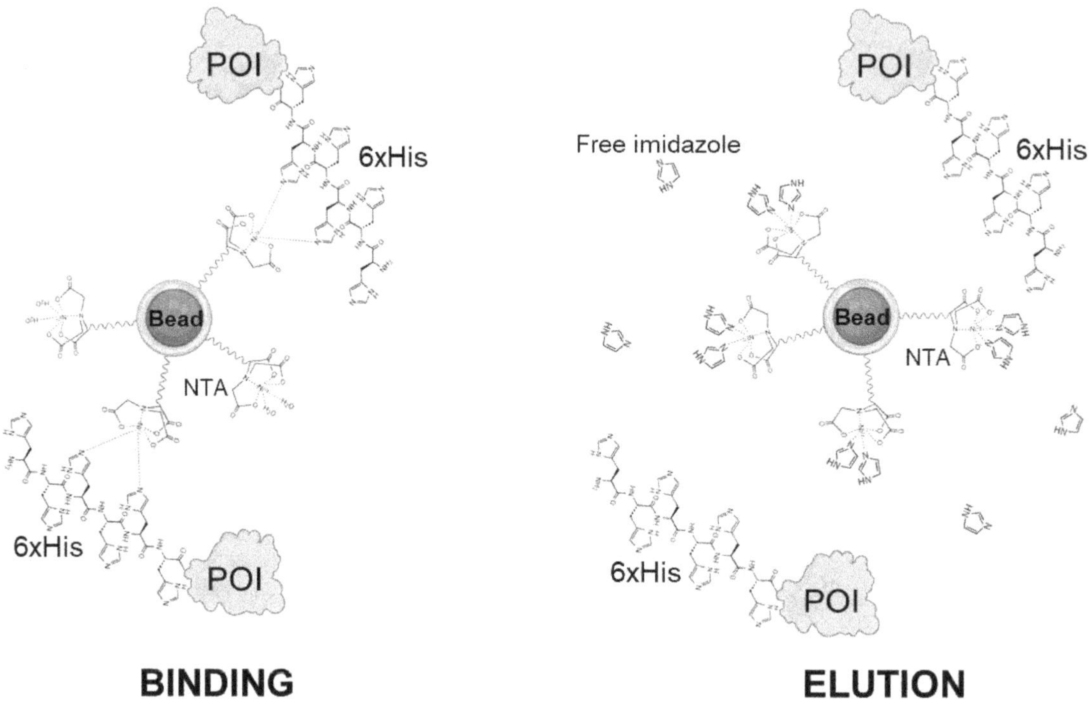

FIGURE 4.5. In nickel affinity chromatography, Ni^{2+} ions are chelated by nitriloacetic acid groups (NTA) linked to stationary phase beads. Pairs of imidazole side chains from the 6His sequence fused to a protein of interest (POI) form coordinate covalent bonds with a Ni^{2+} ion. The cluster of histidines ensures stable, tight binding. After washing away unbound proteins, elution is achieved by adding a high concentration of free imidazole to the mobile phase, which competes with the 6His-tagged POI for Ni^{2+} binding.

PROTEIN TREATMENT AND LONG-TERM STORAGE

Most proteins must be treated very carefully during purification, storage, and use. They are relatively unstable molecules both inside cells and in cell extracts and purified preparations. They are (1) subject to attack by abundant cellular proteases and (2) prone to unfolding or aggregation once purified, which can destroy enzymatic or other biochemical properties.

One of the ways we combat these two problems is to keep proteins as cold as possible during both the purification process and storage. This makes them last a lot longer. Another approach that we will use to combat problem 1 is to add a collection of protease inhibitors to the cell extract prior to the purification procedure. For long-term storage we will transfer the protein to a special buffered solution in which we think it will be fairly happy. One of the important features of this storage solution is a high glycerol concentration. Glycerol has properties that help stabilize proteins in solution and also lowers the freezing point of the protein solution so we can store it at –20°C without it freezing. This keeps the protein solution very cold but prevents the protein from being repeatedly frozen and thawed, which can damage proteins and contribute to problems such as unfolding, aggregation, and loss of activity.

We will use a process called dialysis to transfer the protein from the elution buffer into the storage buffer (Figure 4.6). Dialysis is the selective movement of molecules across a semipermeable membrane. You will place

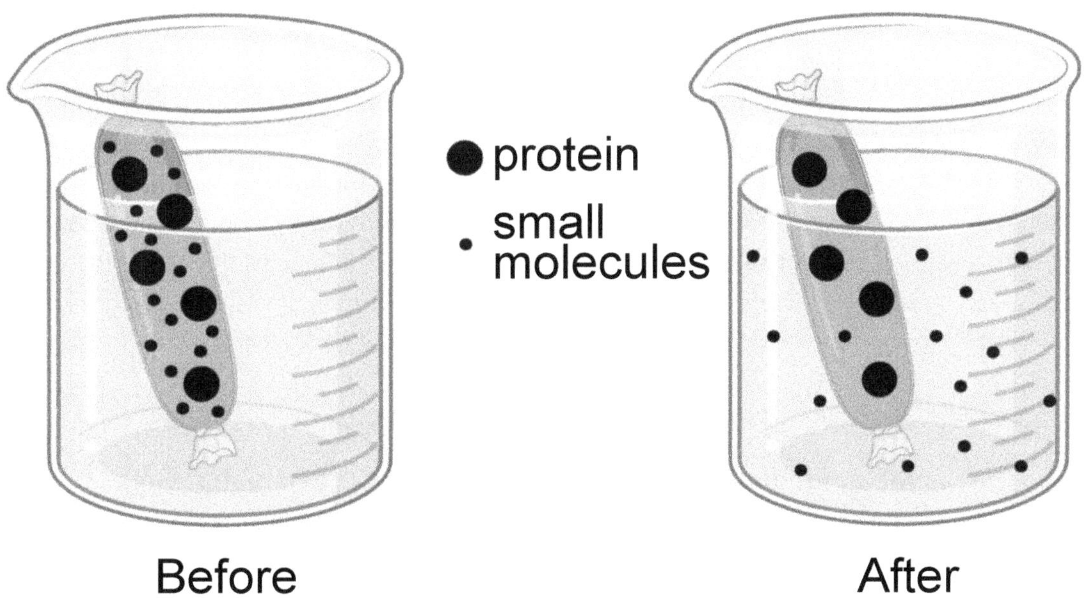

FIGURE 4.6. The process of dialysis for storage of pure proteins. Final protein preparations are sealed in dialysis tubing and placed in a large container of the desired storage solution. Small chemicals freely pass through the membrane and reach equilibrium concentrations on either side. Proteins are too large to pass through the membrane and remain inside the tubing. The process essentially exchanges the initial protein solution for the storage solution, removing unwanted chemicals and achieving the optimal, desired storage conditions.

your small volume of purified protein in a sealed dialysis bag and leave it in a large beaker of storage buffer overnight. Small molecules, such as glycerol and the imidazole we want to get rid of, will pass easily through the pores in the dialysis membrane by diffusion (from high to low concentration until equilibrium is reached), whereas large molecules such as proteins will be trapped in the bag. In the morning the bag will contain your protein in the storage buffer.

MEASURING PROTEIN CONCENTRATION

Knowing the concentration of a purified protein is very important for many downstream studies such as measuring enzyme activity and interactions with other molecules. The Bradford assay is a commonly used method for measuring protein concentration in solution. The procedure was first developed and published in 1976 by Marion Bradford.[1] It uses the dye Coomassie blue to bind proteins. Upon protein binding, the dye turns blue. The absorbance of the resulting solution is measured at 595 nm. Absorbance increases proportionally with protein concentration. A model protein such as bovine serum albumin or gamma globulin is used at known concentrations to create a standard curve representing the relationship between absorbance and protein concentration. This standard curve is then used to determine the unknown concentrations of protein samples simply by measuring their absorbance after incubation with the dye.

OBJECTIVES

The sole objective for the lab today is to purify your target protein, the candidate FoCdc14 protein fused to a 6His tag, using nickel affinity chromatography. You will first generate a soluble protein extract from the pellet of arabinose-induced *E. coli* cells grown last week. Then you will pass the extract through a nickel column, wash away unbound *E. coli* proteins, and elute 6His-FoCdc14 from the column with imidazole. Finally, your protein elution will be dialyzed into a storage buffer, and samples will be prepared for SDS-PAGE analysis of the purification process. You will also learn how to determine protein concentration using the Bradford assay, because this method will be used next week to obtain an accurate measurement of your FoCdc14 concentration.

LAB INSTRUCTIONS

PART 1. PREPARE A SOLUBLE EXTRACT OF *E. COLI* PROTEINS BY CELL LYSIS AND CENTRIFUGATION

1. At the beginning of the morning lab period add 30 ml of cell lysis buffer to your pellet of *E. coli* cells. Use the rubber policeman to resuspend the cells as thoroughly as possible. You don't want to have large chunks of cell pellet remaining—you want a fairly homogeneous suspension.
   Note: A collection of protease inhibitors was added to the lysis buffer you made to help minimize degradation of your protein in the extract.
2. Add lysozyme from the provided 50 mg/ml stock solution to a final concentration of 1 mg/ml. You need to first calculate the correct volume to add.
   Note: Lysozyme is an enzyme that digests bacterial cell walls.
3. Add 4 µl of universal nuclease and mix the suspension well by swirling it. Leave on ice for 30 minutes.
   Note: The universal nuclease degrades RNA and DNA that make the lysate very gooey and viscous.
4. After the 30-minute incubation on ice, add phenylmethylsulfonyl fluoride (PMSF) from the provided 100 mM stock solution to a final concentration of 1 mM. Mix well again and then, using a plastic pipette, transfer the entire lysate to the Oak Ridge centrifuge tube in your ice bucket.
   Note: PMSF is another protease inhibitor. It is added after the cells are lysed because it has a very short lifespan in aqueous solution but it rapidly and irreversibly inhibits serine proteases.
5. Pair up with another group and carefully balance your centrifuge tubes using the provided balances. This is essential to prevent damage to the centrifuge and its rotor. The rotors must be perfectly balanced when spinning at high speed. Add lysis buffer to the lighter tube until they are within 0.1 g of each other.
   Make sure your tube is labeled with your group number.
6. At this point you have a whole cell extract. The extract needs to be clarified to separate insoluble material and cell debris from the soluble proteins. This is done by high-speed centrifugation (30 minutes at 35,000 x g) and will be performed by your teaching assistant in between the morning and afternoon lab periods. The soluble protein extract (the supernatant after centrifugation) will be ready for you when you return to lab in the afternoon.

PART 2. NICKEL AFFINITY PURIFICATION

This protocol is written for ÄKTA Start automated chromatography systems. The ÄKTA Start units from Cytiva are low-pressure peristaltic pump-driven chromatography systems with touch screen control and automated gradient formation, fraction collection, and UV monitoring. An appropriate method that includes column preparation, binding, washing, and elution steps with the 1 ml Ni^{+2}-agarose columns has been programmed already. The two buffer solutions you made last week will act as the mobile phase for your purifications. They both have 25 mM Tris-HCl buffer (pH 7.5), 500 mM NaCl, and 10% glycerol. Buffer B also has 250 mM imidazole. By mixing different ratios of buffers A and B, the system can control the imidazole concentration passing through the column. Make sure you understand what the different buffer components are for and how the chromatography system works. Ask questions. You only need to start the method. However, it is important that you understand what the system is doing and what the different steps in the method are for. This is summarized below.

1. Transfer your 50 ml conical tube containing the soluble *E. coli* protein extract to a beaker filled with ice next to the ÄKTA pump.
   A. Make sure you have at least 32 ml of extract. The method is programmed to load 30 ml onto the column. If you have less, air will get into the column, which may compromise its performance. Increase the volume if necessary by adding some extra lysis buffer. *Record the volume of your extract. You will need this value later for some calculations.*
   B. Transfer 0.2 ml of your clarified extract to a clean microcentrifuge tube. Label this tube "Extract" with your group number and keep this tube on ice.
   C. Cap the conical tube with the cap provided to you with two small holes punched in it. Thread the sample loading line of the ÄKTA through one hole until the tip of the sample tubing is at the bottom of the conical tube. Secure the sample tubing with tape (but don't cover the second hole, as it needs to allow air in as the sample is pumped onto the column).
2. Make sure there is a USB flash drive plugged into the side of the ÄKTA Start unit to save the UV absorbance trace from your purification.
3. Start the ÄKTA method.
   A. On the touch screen select "Method Run."
   B. On the next screen select "User Defined" and then find the method titled BCHM322_Nickel and select it.
   C. Start the run. When prompted for a file name, provide a two-digit number (01 is fine). The extract that flows through the column will be collected in a 50 ml plastic conical tube via the waste line. This has been set up for you.
   D. After ~20 ml of flow-through has been collected, transfer 0.2 ml of this flow-through to a clean microcentrifuge tube. Label this tube "FT" with your group number and keep this tube on ice with the Extract sample.
4. Very important! Fill your fraction collector with 1.5 ml microcentrifuge tubes (inner ring of 30 spots). Caps should be pointed away from the center. Label the tube caps 1–30, corresponding to the numbers on the collector positions.

This is what is programmed in the ÄKTA method:

- 30 ml of extract is loaded onto the column using the sample loading line.
- The column is washed with 20 column volumes (20 ml) of 96% Buffer A/4% Buffer B (25 mM Tris pH 7.5, 500 mM NaCl, 10% glycerol, 10 mM imidazole) using the solvent lines and mixer. This cleans remaining unbound protein out. You should see the UV absorbance trace return to baseline on the graph. Absorbance at 280 nm (UV range) is used to monitor protein level passing out of the column.

- The column is further washed with 7 ml of Buffer with 25 mM imidazole (10% B) to remove contaminating *E. coli* proteins that weakly bind the Ni^{+2}. You should see some material coming off the column with this wash, based on the UV trace.
- A linear gradient of increasing imidazole from 25 mM to 250 mM (10–100% B) is used to elute your target protein from the column. The protein usually comes off near the end of the gradient. Note that the fraction collector will start collecting 1 ml fractions at the beginning of the 25 mM imidazole wash and continue until the gradient is completed. Your protein will be in several of these fractions.
- Finally, the column is washed a bit with 100% buffer B to ensure that all protein is out.
- An image file (.bmp) is saved on the ÄKTA flash drive that you will be able to transfer to your computers and look at. It will allow you to see what fractions should have your target protein, based on the UV absorbance trace throughout the run.

PART 3. USE THE BRADFORD METHOD TO FIND YOUR PEAK OF ELUTED PROTEIN

Although you have a data file that shows what fractions should have your protein, we will confirm this directly by measuring the relative amount of protein in different fractions using the Bradford assay.

1. Perform Bradford assays on your elution fractions.
   A. All assays will consist of 240 µl Bradford dye solution and 10 µl of sample + water. Mix in a microfuge tube and then transfer exactly 200 µl to a microplate well for absorbance measurement at 595 nm on one of the plate readers.
   B. Start by using 10 µl of each fraction directly. Find the peak fraction number in the linear gradient portion of the UV trace by opening and looking at your .bmp data file on your computers. Take samples from this peak fraction and three other fractions before and after the peak, skipping fractions in between. In other words, measure every other fraction, for seven total samples.
   C. Also create a "blank" sample with just 240 µl dye and 10 µl water. You will need this to measure and subtract the background absorbance.
   D. NOTE: If the absorbance value is above ~0.6 after the blank is subtracted, you should reduce the sample volume and try again. The amount needed just depends on how much protein was expressed in the *E. coli* cells and recovered from the nickel column. The linear response range of the Bradford assay is only to ~0.5 Abs_{595}. We don't need to be highly accurate this week, but next week when we measure the exact concentration of your final purified protein we will need to pay attention to the linear range defined by the standard curve.
   E. Create a graph of absorbance versus fraction number in Excel. Does the peak match what is seen in the UV trace?
2. Use the Bradford assay to measure the protein in your original soluble protein extract and flow-through. You may have to make a tenfold dilution of these samples in water and then use a few µl to get an Abs_{595} reading below 0.5. Use the exact same sample setup as above, though. The teaching assistant or the instructor will provide a standard curve that will allow you to calculate the protein concentration in your extract and flow-through samples from the Abs_{595} reading. *Make sure you record all Bradford assay results and calculations. You will need these later.*

PART 4. TRANSFERRING POOLED FRACTIONS INTO STORAGE BUFFER BY DIALYSIS

1. Choose the fractions you want to save for your prep. You want to keep most of the eluted protein peak but don't want fractions with little protein, as it will dilute your concentration. Discuss your selection with the instructor or the teaching assistant for approval. Keep the chosen fractions on ice and discard the remaining fractions.
2. Set up dialysis tubes to transfer your protein into storage buffer.
   A. The teaching assistant or the instructor will cut a 5-inch-long strip of dialysis tubing for you. *Be very careful with the dialysis tubing. Even a small cut or pinhole in the tubing will allow your precious protein to leak out during dialysis, and it will be gone. Handle the dialysis tubing with gloves on.*
   B. The tubing needs to be rinsed with nanopure water before use. The instructor will demonstrate how to open the tubing and rinse it.
   C. Seal one end of the tubing with a dialysis clamp. Fill the tubing completely with nanopure water. Pinch the open end closed and put pressure on the contents of the bag. If the tubing is compromised in any way, you should see water shooting or streaming out. If not, the tubing is okay to use. Empty the water out completely.
3. One group member should hold the top of the tube open. Another group member can then carefully pipet your purified protein fractions into the bag. Seal the top end of the bag with another clamp. Apply pressure again to recheck integrity of the tubing. *Note: The dialysis clamps are prelabeled with your group number, so you don't need to add any labels.*
4. Place your dialysis bag in the large beaker of cold storage buffer. It will be stirred overnight at 4°C. The teaching assistant and/or the instructor will transfer your proteins to microfuge tubes and store them in the -80°C freezer the following morning. You will evaluate the success of your purifications next week.

PART 5. FINAL SAMPLE STORAGE

After completing the Bradford assays, you will need to prepare and save your remaining extract and flow-through samples for later analysis by SDS-PAGE and Western blot.

1. Transfer 100 µl of the extract and flow-through samples to clean microfuge tubes and add 100 µl SDS-PAGE sample dye to each. Heat them at 95°C for two or three minutes.
2. Give the tubes to the teaching assistant or the instructor for storage at -80°C. Make sure they are labeled properly with your group number.

DATA ANALYSIS AND SUBMISSION

Save the following data files and submit to the course learning management system for evaluation:

1. File (.bmp) of the UV absorbance trace from your protein purification on the ÄKTA Start system.
2. An Excel file with (i) absorbance values and graph of absorbance versus elution fraction number, (ii) a list of fractions that were kept and pooled for dialysis, (iii) absorbance readings and volumes used for your extract and flow-through samples, (iv) concentration calculations for your extract and flow-through samples using the provided standard curve.

5. PROTEIN ELECTROPHORESIS

BACKGROUND AND THEORY

ELECTROPHORESIS

Electrophoresis refers to the movement of biological molecules in an electric field. The most common type is gel electrophoresis, which involves the movement of molecules through a solid matrix of polymerized agarose or polyacrylamide in an electric field. Gel electrophoresis is a common and effective method for separating biological macromolecules such as proteins and nucleic acids (RNA and DNA) based on size. Electrophoresis is primarily used as an analytical method for measuring such things as the size, charge, and abundance of molecules after separation. However, electrophoresis can also be used as a preparative method, such as selectively isolating molecules of a particular size.

In today's lab you will use polyacrylamide gel electrophoresis (PAGE) to analyze the protein you purified by affinity chromatography last week. Since proteins are not typically visible to the naked eye in gels, you will use two methods to detect your protein after subjecting it to electrophoresis. One is a general protein stain called Coomassie blue that will allow detection of all proteins present. The other is an immunoblot, or Western blot, that will specifically detect your 6His-tagged fusion proteins (next week).

ELECTROPHORESIS THEORY

The movement of molecules in an electric field depends on their net charge (q), a frictional coefficient (f), and the strength of the electric field (E) and is generally described by the following equation:

$$v \text{ (velocity)} = Eq/f$$

For a given electric field, such as in an electrophoresis experiment, the mobility (μ) of molecules can more simply be described by the equation

$$\mu = q/f$$

The frictional coefficient, f, is influenced by the size and shape of the molecules. If the molecules all have a uniform shape, then f is influenced only by size (mass). In most types of electrophoresis used in biochemical research, the shape of molecules is approximately the same, and therefore mobility depends on the ratio of charge to mass.

The gel matrix provides another means by which molecules are separated according to size. The gel is a molecular sieve, and small molecules can pass through the spaces in the polymerized matrix much more easily than larger molecules and therefore move faster during electrophoresis. Thus, under conditions where shape and charge to mass ratio (or q/f) are constant, the only parameter affecting movement of molecules through a gel is their size (ability to squeeze through the spaces in the gel matrix). This is exactly the basis for denaturing polyacrylamide gel electrophoresis, often called SDS-PAGE. The detergent sodium dodecyl sulfate (SDS) is used to denature (unfold) proteins, giving them a uniform linear shape. SDS is negatively charged, and by binding uniformly to proteins it also gives them a uniform negative charge, independent of the protein's amino acid sequence. Thus, the ratio q/f is identical for all proteins, and the molecular sieve effect of the gel matrix becomes the sole basis for separating proteins, in order of size (Figure 5.1).

POURING POLYACRYLAMIDE GELS

Polyacrylamide gels are created by a chemical reaction that results in polymerization of acrylamide and N, N′-methylene-bis-acrylamide (Figure 5.2). Linear polymers of acrylamide are generated that are cross-linked to each other periodically by the bis-acrylamide units. The size of the pores in the gel matrix can be controlled by varying the concentration of acrylamide and/or the ratio of acrylamide:bis-acrylamide. Typical acrylamide concentrations for separating proteins are 8–12%, with cross-linking ratios from 20:1 to 40:1. Gradient gels in which the acrylamide concentration increases (e.g., from 4–12% or 4–20%) from the top to bottom of the gel are also common and can increase the effective separation range. They require a dedicated gradient-forming

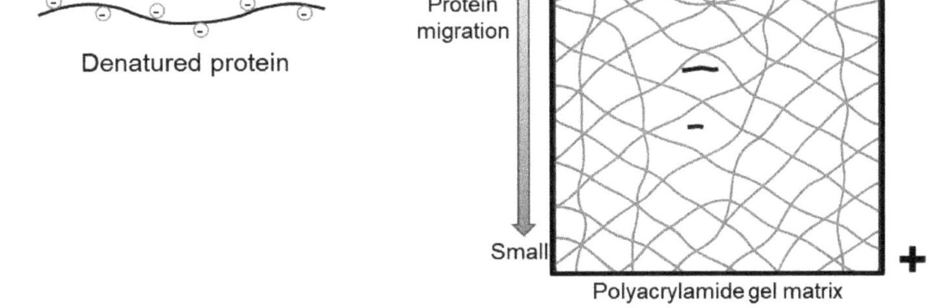

FIGURE 5.1. (A) The ionic detergent SDS coats proteins, causing them to unfold and giving them a uniform charge to mass ratio. (B) In SDS-PAGE the molecular sieve effect of polyacrylamide gels provides the sole means of separating proteins by size during electrophoresis.

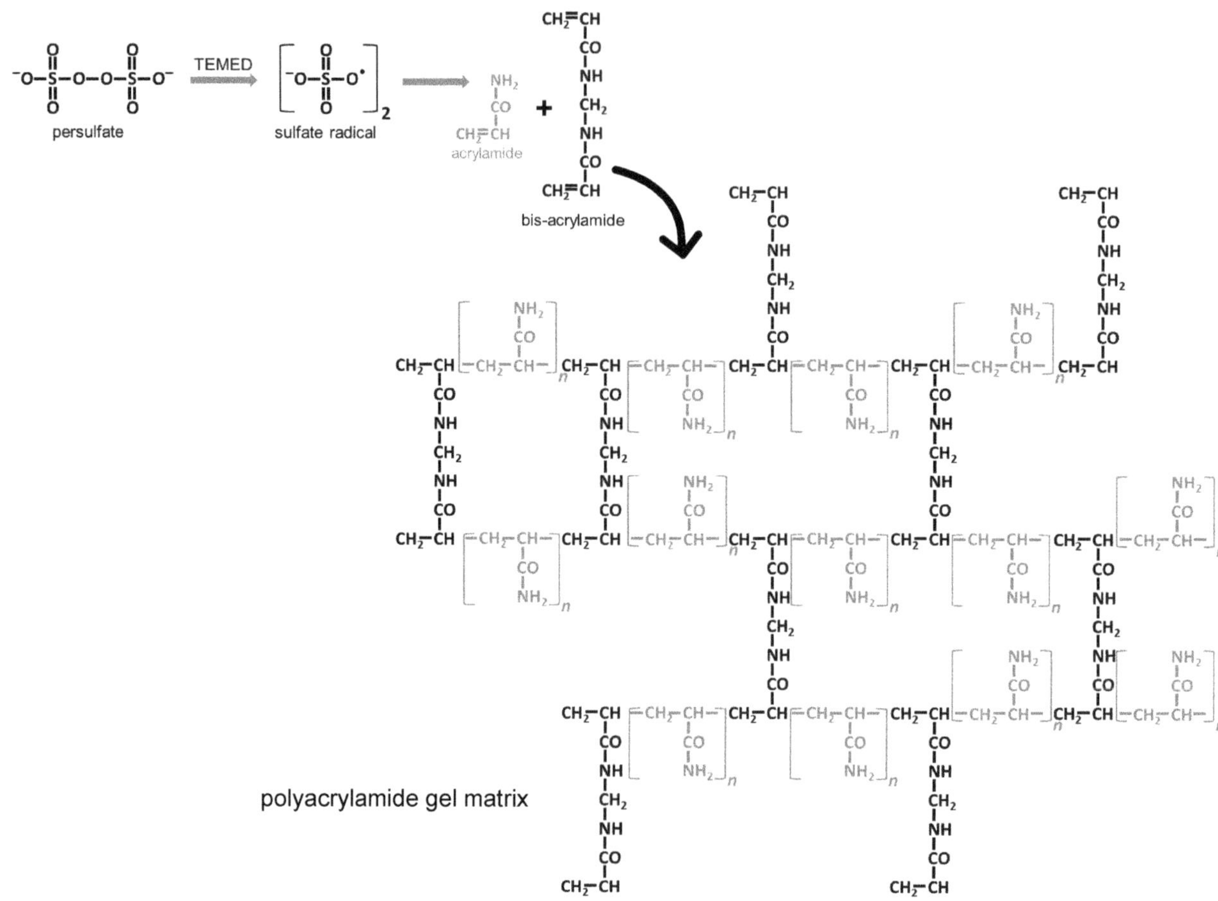

FIGURE 5.2. Radical ions generated by breakdown of persulfate attack acrylamide and bis-acrylamide monomers, activating them to polymerize into a solid 3D mesh. Linear acrylamide polymers are cross-linked by the bifunctional bis-acrylamide units. The resulting gel matrix is the basis for separating denatured proteins by size in SDS-PAGE.

device. The acrylamide polymerization reaction requires free radicals, provided by the chemical ammonium persulfate $(NH_4)_2S_2O_8$ and catalyzed by the compound N,N,N',N'-tetramethylethylenediamine (TEMED).

Many SDS-PAGE systems, including the one you will use, rely on a stacking gel of low pH, low ionic strength, and low acrylamide concentration to initially pack proteins into a very tight band. This band then enters the resolving gel, which has a higher pH and the desired acrylamide concentration for effective protein separation. This system is called discontinuous SDS-PAGE and provides high resolution for protein analysis.

DETERMINING MOLECULAR WEIGHT

Because SDS denatures proteins and makes them linear and coated with a uniform distribution of negative charge, all proteins generally behave the same during SDS-PAGE and are separated strictly based on their size, or molecular weight (MW), as described above. A graph of Log_{10}(MW) versus relative migration distance (or relative front, Rf) through the gel is linear (Figure 5.3). By running a set of standard proteins of known MW

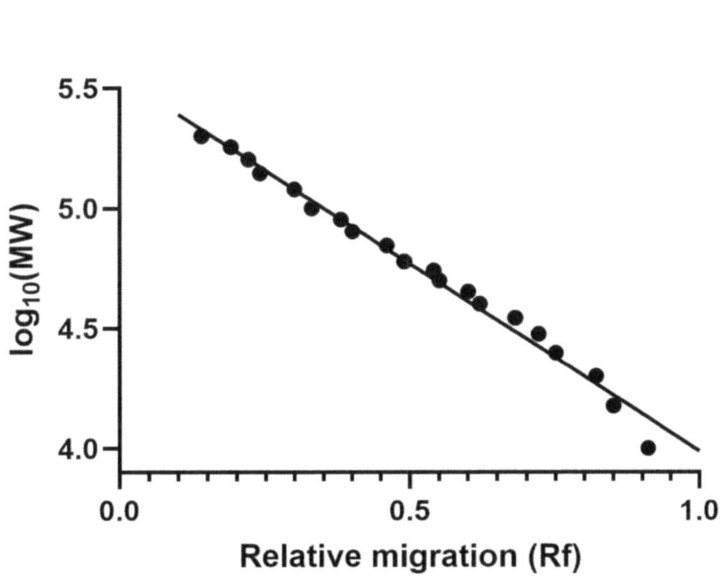

FIGURE 5.3. Proteins migrate in a predictable way during SDS-PAGE. The distance migrated is related to protein length. Specifically, a plot of Log_{10} (protein molecular weight, MW) as a function of migration distance (Rf) is linear, with a negative slope. This allows reasonable estimation of protein MW on an SDS-PAGE gel containing a set of standard proteins of known MW.

on the gel with an experimental sample, a linear standard curve can be generated using this graph format, and the approximate MW of experimental protein(s) can be estimated from their observed migration distance.

VISUALIZING PROTEINS AFTER SDS-PAGE

Most proteins by themselves are invisible when they are separated by electrophoresis. Therefore, after electrophoresis the gels must be stained with a dye that will react with or specifically bind to proteins so that their positions on the gel can be seen. There are a variety of methods for staining proteins in polyacrylamide gels. The most common are

- Coomassie blue, very easy to use, inexpensive, mostly uniform response from proteins (same dye that is used in the Bradford assay for determining protein concentration);
- Reduced silver, a more complicated and technically difficult procedure with response that is protein-dependent and with the advantage of being extremely sensitive; and
- Fluorescent dyes, an easy to use option for accurate quantitation of protein abundance but comparatively expensive and requires specialized imagers for visualization.

OBJECTIVES

Objective 1 for today's lab is to evaluate the success of the recombinant 6His protein expression and purification activities from the past few weeks. This will be achieved by SDS-PAGE, in which protein samples from the uninduced and induced *E. coli* cells, the soluble protein extract and Ni^{2+} column flow-through, and final eluted, dialyzed protein will be separated and stained with the general protein dye Coomassie blue. While the gels will not be ready to view until next week, these gels should provide information on success of the arabinose-induced expression system, Ni^{2+} binding, and target protein recovery and purity. In addition to the Coomassie blue-stained gel, proteins from a second gel will be transferred to a polyvinylidene difluoride (PVDF) membrane for immunoblotting (Western blotting), which will be the main subject of next week's lab.

Objective 2 for today's lab is to measure the solution concentration of your final purified protein after last week's dialysis into storage buffer. Concentration is critically important in many biochemical procedures, and you will need to use this protein concentration multiple times throughout the project. It is important to have an accurate, reliable measurement. You were introduced to the Bradford assay last week and the concept of standard curves. This week you will create your own standard curve for the Bradford assay and learn how to perform careful concentration measurements with it.

LAB INSTRUCTIONS

PART 1. POURING GELS

Lab safety note: Liquid acrylamide is one of the few chemicals we use that is dangerous. It is toxic and readily absorbed through the skin. It is critically important to wear gloves, lab coats, and protective eyewear when working with the liquid acrylamide.

In the morning lab period,

1. Set up the gel casting stands and two sets of plates as demonstrated by the instructor.
2. Mix the following reagents together to create the resolving gel in a 50 ml conical tube:
   A. 6.4 ml water
   B. 4 ml resolving gel buffer (1.5 M Tris-HCl pH 8.8)
   C. 5.3 ml 30% acrylamide/bis-acrylamide solution (29:1)
   D. 160 µl 10% sodium dodecyl sulfate (SDS)
   E. 160 µl 10% ammonium persulfate (APS)
   F. 40 µl TEMED (add last)
3. Mix well and then immediately add the gel mixture to the space between the short and tall glass plates using a disposable transfer pipet. Bring the level up to the top of the green clamp arms on the gel assembly. Cover the surface with a thin layer of isopropanol (using a transfer pipet) and allow to harden for ~20 minutes. Use the transfer pipet and remaining liquid in the conical tube to determine when the acrylamide has polymerized. Pour off alcohol in the sink and blot dry with a Kimwipe.
4. While waiting, mix the following reagents together to create the stacking gel (but **DO NOT** add APS and TEMED until ready to pour when the resolving gel has hardened):
   A. 4.9 ml water
   B. 2 ml stacking gel buffer (0.5 M Tris-HCl pH 6.8)
   C. 1 ml 30% acrylamide/bis-acrylamide solution
   D. 80 µl 10% SDS
   E. 80 µl APS and 20 µl TEMED (add last, only after the resolving gel has polymerized and you are ready to add the stacking gel mixture to the plates)

5. With another transfer pipet, add the stacking gel mixture on top of the hardened resolving gel up to the top of the short glass plates. Insert 10-well plastic combs into the two gel assemblies, making sure no bubbles are trapped, and allow to harden until the afternoon lab session.

PART 2. SAMPLE PREPARATION

1. You will be running the following samples on your gels:
   A. Your pre- and postinduction cell extract samples from Week 3—already prepared in SDS sample dye. Nothing else needs to be done.
   B. Your extract and flow-through samples from Week 4—already in SDS sample dye. Nothing else needs to be done. You calculated the protein concentration of these samples last week.
   C. Your final purified protein from Week 4.
2. Determine the concentration of your final purified protein.
   A. Use the microplate spectrophotometers and the same Bradford assay protocol used last week to create your own standard curve. For every standard, mix 240 µl of Bradford dye and 10 µl of sample (protein + water) in a microfuge tube by vortexing. Transfer 200 µl of this mixture to a microplate well for absorbance measurement at 595 nm on the plate reader.
   B. Measure the absorbance of 0, 0.25, 0.5, 1, 2, 4, 6, and 8 µg of standard protein using the provided series of standard stock solutions. The stocks are set up so you can just add 10 µl directly to the 240 µl of Bradford dye, then transfer 200 µl for absorbance measurement in the microplate. Use 10 µl water for the blank sample with 0 µg protein.
   C. After measuring all standard samples, transfer the data to an Excel file. Be sure to subtract the blank (0 µg standard) from all samples and generate a graph of absorbance versus µg protein. Include your blank sample as a 0,0 data point. Perform linear regression, force the line through the origin, and include the resulting equation on the graph. With this equation, or "standard curve," you can calculate the protein concentration in your samples from similar absorbance measurements.
   D. Measuring your pure protein: Get 2 different Abs_{595} readings within your standard curve range and calculate the protein concentration from them. Perform each in triplicate and get an average and standard deviation for your total protein concentration. Make sure that the two different amounts give similar concentration values. All protein preps have different concentrations, so you will need to determine the appropriate volume to use by trial and error. You can start with 2 µl, but for some preps you may need to make a tenfold dilution first so you are pipetting a reasonable volume. If you have to use less than 2 µl, then a dilution is best to get accurate measurements (so you aren't trying to transfer a tiny volume).
3. Determine the volume of your protein sample needed to get 5, 2, and 1 µg (again, a dilution may be required). Once you have finished that calculation, mix the following components:
   A. 5 µl 4x SDS sample dye + x µl protein sample (to get either 5, 2, or 1 µg) + (15-x) µl water.
   B. Total volume should be 20 µl. You will load all of these samples onto the gel for Coomassie blue staining.
4. Determine the volume of your protein sample needed to get 0.5 µg. Make a similar mixture for gel loading as in step 3 (volume = 20 µl). You will split this sample and load 4 µl and 16 µl into different lanes on the gel for immunoblotting.
5. Calculate the amount of your extract and flow-through samples needed to get 50 µg loaded on the gel based on your concentration measurements. Keep in mind that you mixed your samples 1:1 with SDS sample dye last week, so the effective concentration is now half of what you measured by Bradford.

PART 3. RUNNING THE GELS

1. Before loading your gels, heat all samples at 95°C for 2 minutes and spin briefly (a few seconds) in a microfuge.
2. For the Coomassie blue gel:
   A. Load 8 µl of the unstained molecular weight markers in lane 1. The teaching assistant or the instructor will demonstrate how to properly load samples into the wells. Try to avoid bleed-over of sample to adjacent lanes, as this can confuse interpretation of the results.
   B. Load 20 µl of your pre- and postinduction cell extract samples from Week 3 in lanes 2 and 3.
   C. Skip a lane and load the required volume for 50 µg of the extract and flow-through samples in lanes 5 and 6.
   D. Skip another lane and load the entire volume of your purified protein samples in lanes 8–10 (5, 2, and 1 µg).
3. For the immunoblot gel:
   A. Load 5 µl of the prestained molecular weight markers in lane 1.
   B. Load 10 µl of your pre- and postinduction cell extract samples from Week 3 in lanes 2 and 3 (this is half the amount loaded for the Coomassie gel).
   C. Load 25 µg of your protein extract and flow-through samples in lanes 4 and 5 (again, this is half the volume used for the Coomassie gel).
   D. Skip a lane and load 4 µl of your 0.5 µg pure protein prep in lane 7.
   E. Load the remaining 16 µl of your 0.5 µg pure protein prep in lane 8.
4. Finish assembly of the running apparatus (the teaching assistant or the instructor will show you how to do this) and turn on the power supply. Run the gels at 180 V for 1 hour or until the dye front reaches the bottom of the gel.
5. Disassemble the gel apparatus and use a green plastic gel knife to pry the plates apart to access the gel (it should stick to one plate). The teaching assistant or the instructor will help with this.

PART 4. STAINING

Wear gloves when handling the gels.

1. Transfer the gel for Coomassie blue staining to a Tupperware container with some nanopure water in it. Wash the gel twice with water for ~5 minutes each.
2. Discard the second water wash in the sink and cover with the provided GelCode Blue safe Coomassie blue dye solution.
3. Incubate on the gel rocker.
4. The gels will stain overnight. The following day the teaching assistant will remove the Coomassie solution and replace it with water. The stained gels will be stored for imaging and analysis next week.

PART 5. TRANSFER TO PVDF

Wear gloves when handling the PVDF and gels.

1. Prepare a piece of PVDF membrane by soaking it in a tray of methanol for 10 seconds and then immediately moving it with forceps to a Tupperware container filled with transfer buffer. Ensure that the PVDF remains completely submerged. Leave it in the transfer buffer for at least five minutes.

2. Carefully lay the PVDF membrane over the gel, ensuring that no bubbles exist between them. You can use a plastic pipet as a roller to remove any bubbles (the gel knife can be used for this purpose as well to push bubbles out toward the edge of the gel). NOTE: It is critical that no bubbles exist between the membrane and gel or you will have blank spots in your immunoblot.
3. Wet a stack of blotting paper in transfer buffer and lay it on the bottom electrode plate in the Transblot Turbo cassette. Carefully transfer the membrane and the gel onto the blotting paper with the PVDF on the bottom. The gel must be on top of the PVDF, because electrophoretic transfer will be from the top electrode to the bottom electrode plate.
4. Finally, soak a second stack of blotting paper in transfer buffer and place them carefully over the gel. Make sure the stack is aligned so that the gel is completely covered. Transfer a couple extra ml of transfer buffer onto the gel/membrane sandwich to ensure that it is fully wet and then seal the cassette with the top electrode plate (the teaching assistant or the instructor will do this).
5. The teaching assistant or the instructor will start the method for rapid transfer of the proteins out of the gel and onto the membrane.
6. Disassemble the sandwiches. First, confirm that the prestained markers have mostly transferred to the membrane by peeling a corner of the gel off the membrane. If transfer is good, carefully move membranes to a square blotting tray containing some nanopure water using membrane forceps. Make sure the protein side is up—that is, the side that was touching the gel. Discard the blotting paper and the gel. Label the tray lid with your group number. Wash the membrane twice with water for a couple of minutes each and then discard all the water. Membranes will be air-dried and stored until next week for the immunoblot analysis.

   Note: The night before the next lab the teaching assistant will rewet the membranes and incubate them in blocking solution, which is the PBS you made two weeks ago, supplemented with the detergent Tween-20 and 5% nonfat dry milk. The milk protein casein will stick to the PVDF uniformly to prevent your antibody from sticking to it. This ensures that the antibody only binds to its antigen, not to the PVDF.

DATA ANALYSIS AND SUBMISSION

Submit the following data as a single spreadsheet file to the course learning management systems (and be sure to save to your group's designated lab data storage location as well):

1. Bradford standard curve with linear regression.
2. Calculations of your purified protein concentration (include individual calculations and the average and standard deviations) and total amount of protein in your final prep. Be sure to include all work using Excel functions so that the instructor and the teaching assistant can see how you did the calculations.

6. IMMUNOBLOTTING

BACKGROUND AND THEORY

THE IMMUNOBLOTTING METHOD

Immunoblotting (also called Western blotting) is a method to detect specific proteins of interest within a complex mixture of proteins after separation by SDS-PAGE. Immunoblotting provides very high specificity and sensitivity. General protein stains such as Coomassie blue detect all proteins present in a sample indiscriminately but provide no information about the identity of any of the proteins. Immunoblotting allows one to determine if a specific protein of interest is present or not and, with the appropriate controls, how much of it is present.

In order to perform an immunoblot, proteins that have been separated by SDS-PAGE must be transferred from inside the gel matrix onto the surface of a membrane so they are exposed and can be probed and detected. Transfer is done electrophoretically. A membrane, usually made of nitrocellulose or polyvinylidine difluoride (PVDF), is placed in direct contact with the gel surface. The membrane material binds proteins very tightly. Filter paper is layered on both sides, and the "sandwich" is placed between two metal electrodes in a transfer apparatus, either a semidry unit, as shown in Figure 6.1, or a fully submerged tank system. An electrical current

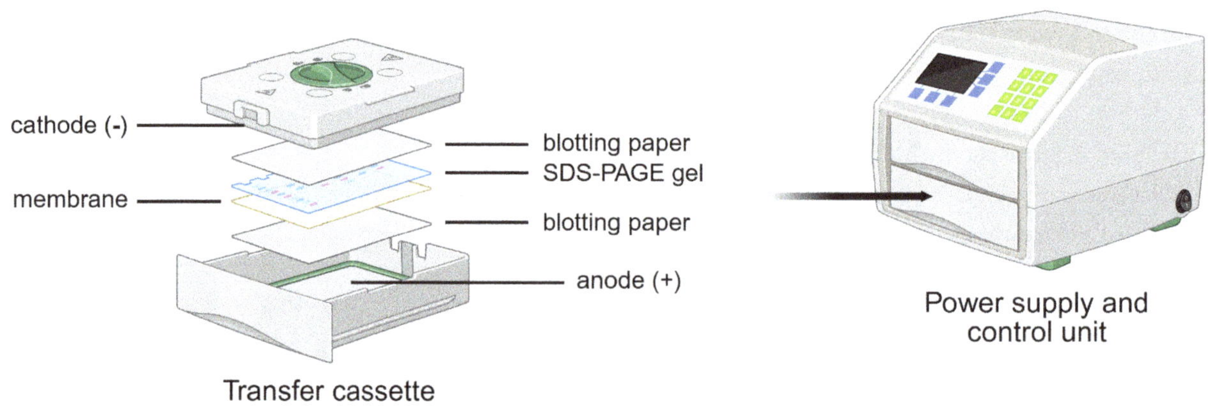

FIGURE 6.1. For immunoblotting, proteins are electrophoretically transferred onto a solid membrane after separation by SDS-PAGE. The membrane is placed in contact with the gel, with blotting paper on either side soaked in an appropriate electrophoresis buffer. This stack is sandwiched between two electrode plates that generate current and mediate migration of proteins out of the gel onto the surface of the membrane. As in SDS-PAGE, proteins move toward the positive anode. Created in BioRender. Hall, M. (2026) https://BioRender.com/b5rysxo.

is then established between the electrodes, and proteins migrate horizontally out of the gel toward the positive electrode and stick to the surface of the membrane.

Once transfer is complete, all remaining protein binding sites on the membrane must be blocked. This is done by soaking the membrane in a solution of a nonspecific protein such as bovine serum albumin (BSA) or casein (i.e., milk). This coats the membrane with protein and prevents the antibodies used in later steps from sticking all over the membrane.

Next, the membrane is incubated with a solution of primary antibody. The primary antibody specifically recognizes the protein of interest. Today you will be using a mouse antibody that binds specifically to the 6x histidine tag on your 6His-FoCdc14 fusion protein. Antibody-antigen interactions are incredibly strong and will persist despite extensive washing steps. You learned about where antibodies come from, what they look like, how they are generated for specific target antigens of interest, and the breadth of applications they can be used for in the lecture associated with this week's lab. The anti-6His antibody is useful because it can be used to detect any protein being expressed for purification via Ni^{2+} affinity chromatography. Since FoCdc14 has never been characterized before, it is not surprising that no commercial antibodies exist to recognize it.

In most immunoblotting protocols (Figure 6.2), after washing away excess primary antibody the membrane is incubated with a solution of secondary antibody. The secondary antibody binds tightly to any antibody from the species in which the primary antibody was raised. For example, if the primary antibody was raised in rabbits, then it consists of rabbit immunoglobulin protein. Thus, an anti-rabbit immunoglobulin secondary antibody would be used. The secondary antibody will bind very tightly to the primary antibody wherever it is located on the membrane. The secondary antibody is also covalently linked to a molecule that will allow visual detection of its location. This molecule is usually either an enzyme called horseradish peroxidase (HRP) or a fluorescent molecule. Excess secondary antibody must also be washed away prior to imaging the blot. In the protocol for this lab, you are using a shortcut. Since the anti-6His antibody is so commonly used, companies have generated anti-6His antibody-HRP fusions that eliminate the need for a secondary antibody. Thus, the detection system is linked directly to the primary antibody. This makes the protocol shorter and simpler.

In the final step, the membrane is incubated with a solution containing a substrate of HRP. This can be either a substrate that is converted to a colored insoluble product for immediate visual detection or a light-generating compound, luminol. The oxidized product of luminol, 3-aminophthalate, emits light (chemiluminescence) very briefly before it can diffuse away from the reaction site. The emitted light allows the antigen band to be detected by specialized light-detecting imagers, whereas all the other proteins on the membrane remain invisible. You will use this chemiluminescent detection system today.

If a colorimetric (color-forming) substrate is used, no further processing is required. If a chemiluminescent substrate is used (or if fluorescent secondary antibodies are used rather than an HRP conjugate), the blot must be exposed to photographic or radiographic film in a darkroom (for chemiluminescence) or an imaging device that can detect emitted light. Chemiluminescence and fluorescence are more sensitive detection methods. You will be using the department's digital imaging system to visualize the chemiluminescent signal on your immunoblots today. The supercooled CCD camera in this imaging system provides superior quantitative capacity compared to film because of its large dynamic range. Film has a very narrow dynamic range. Dynamic range refers to the range of signal intensities that can be detected.

It is important to understand that the amount of signal detected in an immunoblot is not sufficient to determine how much of the antigen is present. This is because every antibody is unique, and there can be high

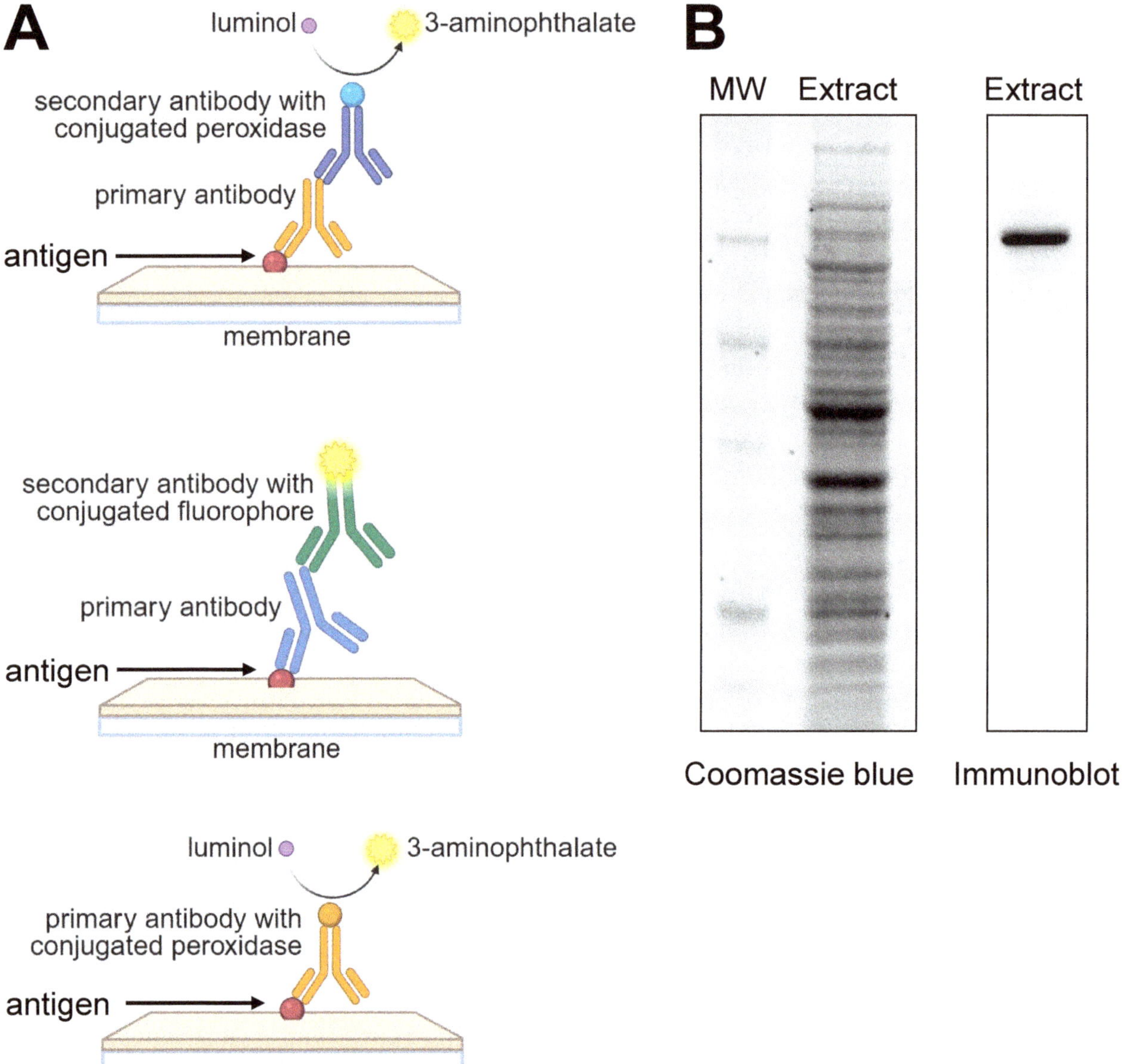

FIGURE 6.2. (A) Three common antigen detection strategies in immunoblotting. *Top*: A secondary antibody is conjugated to a peroxidase that converts luminol into the luminescent product 3-aminopthalate in the presence of hydrogen peroxide. The secondary antibody recognizes antibodies from the host species used to generate the primary antibody. *Middle*: The secondary antibody is conjugated to a fluorophore for detection of fluorescent light. *Bottom*: The primary antibody is conjugated to a fluorophore or peroxidase enzyme, bypassing the need for a secondary antibody. (B) The power of immunoblotting is shown by comparing a protein extract separated by SDS-PAGE and then either stained with Coomassie blue dye to reveal all proteins or probed with an antibody for specific detection of a single protein of interest. MW, molecular weight markers. Created in BioRender. Hall, M. (2026) https://BioRender.com/s4qapxp.

variability in the blotting and detection process. However, within a single blot, the relative amount of antigen in different samples can be compared to determine how its abundance changes in response to experimental variables. In addition, if known amounts of pure antigen are available and included on the blot, then the absolute amount of antigen in experimental samples can be calculated from these standards (much like the use of a protein standard curve with the Bradford assay last week).

FOLD ENRICHMENT AND PERCENT YIELD IN PROTEIN PURIFICATION

It is useful to have measurable values that reflect the effectiveness of protein purification procedures. The higher the purity, the less likely other contaminating proteins will contribute to observed results. Yield may be important if a large amount of protein is needed to conduct experimental procedures. "Fold enrichment" is a classical biochemical term reflecting the increase in a protein's purity during the course of isolating it from its biological source. The term simply refers to the increase in the fraction of total protein attributable to the target protein being purified. For example, a target protein may represent only 0.001% of total protein in a cell, but after the purification procedure it represents 80% of the total protein isolated. The fold enrichment is therefore 80 ÷ 0.001, or 80,000 fold. You could say that the isolated protein is 80,000 fold more pure at the end of the procedure than it was at the beginning. The fold enrichment, or fold purification, is not an indication of the actual purity of the final sample, just the change in purity over the course of the procedure.

Percent yield reflects the amount of the target protein that is recovered at the end of the purification procedure. If a biological sample such as an extract from cell culture contains 10 mg of the target protein and the final purified protein sample has 2 mg, then the percent yield is 20%. Every step of a purification has some associated sample loss that reduces yield. It is common in biochemistry to report the fold enrichment and the yield of each step in the purification process (Table 6.1). The values can be useful when, for example, comparing different purification strategies to find the best option for a protein of interest.

OBJECTIVES

The primary objective of today's lab, with respect to the research projects, is to characterize the success of the FoCdc14 purification by measuring the fold enrichment and percent yield, two commonly measured indicators of protein purification from a biological source. These values will be determined from an immunoblotting

TABLE 6.1 *Protein Purification Table*

PURIFICATION STEP	TOTAL PROTEIN (MG)	SPECIFIC ACTIVITY (UNITS/MG)	PERCENT YIELD (%)	FOLD ENRICHMENT
Crude cell extract	10,000	5	100	1.0
$(NH_4)_2SO_4$ precipitation	300	120	72	24
Ion-exchange chromatography	80	350	56	70
Affinity chromatography	12	1,670	40	334
Size-Exclusion chromatography	6	3,000	36	600

experiment, and this provides a useful opportunity for students to learn what antibodies are and how they are used in research applications.

A secondary objective is to evaluate the Coomassie blue-stained protein gels from last week and then excise the putative purified FoCdc14 band and process it for mass spectrometry analysis. Mass spectrometry will be used to test if this band is the expected protein. In doing so, students will learn about how mass spectrometry data can be used to identify proteins present in biological samples, an important part of the field of proteomics.

LAB INSTRUCTIONS

PART 1. IMMUNOBLOTTING

1. Transfer of proteins to a PVDF membrane was completed last week. The membranes have been incubated in blocking solution for you to coat all other protein binding sites. Blocking solution is the PBS you made in Week 3 with the detergent Tween-20 (PBS-T) and 5% nonfat dried milk (the milk protein casein serves as the nonspecific blocking agent). The membranes are now sitting in PBS-T.
2. Carefully pour off the PBS-T into the sink and add primary antibody solution to the membrane. To make the primary antibody solution, add 4 µl of the anti-6His antibody-HRP conjugate to 8 ml PBS-T + 1% milk. Mix well by inverting the tube several times. This is a 1:2,000 dilution. Place the tray with your membrane and antibody solution on one of the rockers. Incubation with primary antibody will proceed until the afternoon lab period begins.
3. Carefully pour off the primary antibody solution into the sink at the beginning of the afternoon lab period. Quickly rinse it off with some PBS-T. Then wash the membrane three times for 10 minutes each on the rocker with PBS-T (~10–15 ml per wash) to remove residual antibody solution.
4. Pour off the final PBS-T wash. Apply 2 ml of detection reagent to your membranes. The detection reagent contains the luminol, which becomes chemiluminescent after oxidation by the horseradish peroxidase (coupled to the primary antibody) in the presence of hydrogen peroxide. Keep the membrane evenly coated by manually rocking the tray and proceed to the digital imager system with the instructor or the teaching assistant.
5. The instructor or the teaching assistant will work with each group on the Bio-Rad ChemiDoc MP imaging system to collect an image of their blot and save it to the group's flash drive. Data files should then be transferred to your computers and data storage folders.

PART 2. MASS SPECTROMETRY SAMPLE PREPARATION

In the morning prep period:

1. Acquire a digital image of your Coomassie blue-stained gel from last week with the help of the teaching assistant or the instructor. Save the file to your flash drive and transfer it to your computers and data storage folder.
2. Back in the lab, use a razor blade to carefully cut out the band you think represents your purified protein. Be sure to wear gloves when handling the gel.
3. Cut the gel band into four or five smaller pieces using a spatula and a razor blade and transfer the pieces into a provided 1.5 ml microfuge tube. Add 100 µl HPLC-grade water.
4. Label the tube with your group number, close the cap, and incubate for 15 minutes on the bench, vortexing it periodically.

5. In the meantime, make a solution of 50% acetonitrile in 25 mM ammonium bicarbonate by mixing 500 μl each of 100% acetonitrile and 50 mM ammonium bicarbonate in a 1.5 ml microfuge tube. This is the gel destaining solution.
6. Remove the water from your gel pieces with a micropipettor and cover the pieces with 400 μl of the gel destaining solution. Vortex briefly.
7. Leave the labeled tubes in the rack on your bench. The destained pieces will be ready to digest with trypsin in the afternoon.

In the afternoon lab period:

1. Remove as much of the destaining solution as possible with a micropipettor. Add 100 μl 100% acetonitrile to the tube and let it sit on the bench for 10 minutes. Remove the solution and repeat with another 100 μl acetonitrile. After 10 minutes, the gel slices should be small, white, and hard. You can check this by touching them with a pipet tip. If still squishy, repeat the acetonitrile wash again.
2. Remove as much acetonitrile as possible from the dried gel pieces. Add 40 μl of trypsin solution (20 μg/ml in fresh 50 mM ammonium bicarbonate) to the gel pieces. The gel pieces will absorb the trypsin solution, and the proteins will be digested inside the gel.
3. The tubes will be incubated at 37°C overnight. Tomorrow the instructor or the teaching assistant will extract the resulting peptides from the slices with acetonitrile to perform mass spectrometry analysis on.

PART 3. CALCULATING MOLECULAR WEIGHT, PURITY, % YIELD, AND FOLD ENRICHMENT

Molecular weight (MW) estimation:

1. Use the pI/MW tool from the Expasy website that you used previously in Week 1 (https://web.expasy.org/compute_pi/) to calculate the predicted MW of the 6His-FoCdc14 protein you expressed in *E. coli*. You have the sequence for *6His-FoCDC14* in the SnapGene file used for DNA sequencing analysis in Week 3 (remember, the sequence used for expression is shorter than the full *FoCDC14* gene sequence analyzed in Week 1 and has been codon-optimized for *E. coli* expression). Highlight the DNA coding sequence and then copy the amino acid sequence from the Edit menu.
2. Open the image file of your Coomassie blue-stained gel in ImageLab on your computer.
3. Use the "Lanes and Bands" toolbar button at left to define the gel lanes using the manual lane finder function. The instructor will go through this with you. The top of the lanes should be the boundary between the stacking and resolving gels. The bottom of the lanes should be above the dye front, or the bottom of the gel if you ran the dye off.
4. Next, select the "Bands" tab. It is easy to manually select the bands for which you want to determine the MW. Simply click the "Add" button and then click on the band for your protein in each of the lanes in which you can visibly see it (hopefully all of them). If there is uncertainty about which band is your target protein, ask the teaching assistant or the instructor for help.
5. Now, go back up one level to the main toolbar menu and select the "MW Analysis Tools" button. In the Standards window select the "New" button. Name the new marker set—something like "Group # MW Markers" is fine. Click the add button and type the kDa value for each MW standard protein—these values will be provided by the instructor. Then, check the box underneath the gel lane corresponding to the MW markers. This should be Lane 1 if you

loaded the samples as instructed. You should see that the protein marker bands are automatically detected. If not, you can manually delete or add bands as needed by going back to "Lanes and Bands."

6. View your MW standard curve by clicking on the "Standard Curve" icon in the menu bar. It should be linear when the Log Y-axis box is checked if your standards are labeled correctly. Save a screenshot of this standard curve to include in your data submission.
7. All you have to do to find the estimated MW of your protein now is click on the Analysis Table menu item at the top of the screen and scroll over until you see the lanes in which you selected your target protein band. Record the MW value and save a screenshot of the Analysis Table showing the result. Is your purified protein the expected size?
8. Use the Annotations tool in ImageLab to label all of the lanes on your gel, and save this modified ImageLab file to your data storage folder. You will need to make a nice figure of your purification gel with appropriate lane labels for your final report.

Estimating FoCdc14 purity:

1. When you select a band within a lane, the ImageLab program automatically calculates the percentage of the total signal in that lane contributed by that band. This is a crude measure of the purity of the protein. This value can again be found in the Analysis Table. Record the % purity of your protein from each lane. Make sure that your lane does not include signal from the sample dye, as that signal will count against the FoCdc14 purity.

Using immunoblot data to estimate fold enrichment of FoCdc14:

The extent of purification, often called fold enrichment or fold purification, can be obtained by comparing the abundance of your target protein in the initial extract and the final stored preparation relative to total protein.

1. Open your immunoblot image file in the ImageLab software. Label the lanes appropriately using the Annotation tool, and save the labeled image to your lab data storage location.
2. Using the ImageLab software as described above for the Coomassie gel, determine the integrated volume of each immunoblot band (extract, flow-through, and final pure protein). This just requires selecting each band you want to quantify and then finding the volume value in the analysis table—it will be the column labeled "Adj. Volume (Int)."
3. You have total protein concentration measurements of your initial extract, flow-through, and purified protein based on Bradford assays.
4. Using just the band intensities (Adj. Volume (Int)), the protein concentrations, and the volumes of each sample that you loaded in the gel lanes, you should be able to calculate the fold enrichment achieved in your purification. To do this calculation you need only calculate the band intensity/µg protein loaded for both the starting extract and the final pure protein prep. Record these values.

Hint: What you are looking for is the fraction of total protein in the extract and the final dialyzed fraction that is FoCdc14. Since you can't measure the amount of FoCdc14 directly from the immunoblot, you have to use the band intensity as a surrogate. You want to determine the signal intensity per µg protein loaded for both the extract and the final pure protein. The fold enrichment will then be a ratio of these values (final/extract).

Using immunoblot data to estimate percent yield from FoCdc14 purification:

Comparing the abundance of your protein in the extract and flow-through can provide an estimate of the % yield. Yield can also be estimated by comparing the extract to the final purified protein, provided that the volume loaded on the gel and the total volume of each sample are known. The instructor will give you the volume of your final protein sample. You know the volume of your initial extract (~32 ml).

1. Simple estimate. Dividing the flow-through intensity by the extract intensity and multiplying by 100 will reveal the % protein "lost" that didn't bind to the column. Subtracting this from 100 will provide the protein yield in %—that is, the % of the protein in the extract that was retained on the Ni2+-agarose column.
2. The simple estimate in step 1 makes the assumption that everything lost as the original extract was passed through the Ni2+ column ended up in our final stored protein preps. A more accurate approach to measure yield is to compare the amount of final purified FoCdc14 to the amount of FoCdc14 in the starting extract. You can do this using the band intensity values originally obtained above, but the calculation is trickier. See if you can figure out how to calculate the % yield this way.

Hint: For % yield in step 2, the other values you need to use are the volume of each sample (extract and pure protein) loaded on the gel, and the total volume of those samples.

DATA ANALYSIS AND SUBMISSION

Submit a single Excel spreadsheet file containing the following data and calculations to the course learning management system (including all work for the calculations using Excel functions). In the Excel file create and name individual sheets for each of the following calculations so it is easy to follow:

1. Estimated MW of target protein from SDS-PAGE image analysis (make sure your ImageLab file is saved in your data storage location).
   A. Include an image of your gel and a screenshot of both the MW standard curve and the Analysis Table showing the measured MW for your purified protein.
   B. Using the MW analysis and the uninduced and induced extract lanes from your gel, describe the existing evidence that you have successfully purified the intended FoCdc14 protein.
2. Estimated % purity from ImageLab analysis of each lane of purified protein. You can simply copy the analysis table from ImageLab containing the lane purity values.
3. Your fold enrichment calculations showing all work. This page should include an image of your Western blot with labeled lanes.
4. Your % yield calculations (using both approaches) showing all work.

7. PROTEIN MASS SPECTROMETRY

BACKGROUND AND THEORY

MASS SPECTROMETRY AND ITS APPLICATION TO PROTEINS

Until the mid-1990s, it was very difficult to identify or determine the amino acid sequence of an unknown protein. That changed with the development of two seemingly unrelated technologies. First, high-throughput DNA sequencers began sequencing the entire genomes of various species, starting with the simple bacterium *Haemophilus influenzae* and the eukaryotic fungus *Saccharomyces cerevisiae*. Having an entire genome sequence allows prediction of all possible protein sequences encoded by that genome. Now, thousands of organisms have had their genomes sequenced, including humans. Second, new mass spectrometry methods were developed for the sensitive and routine analysis of proteins and peptides. Previously, mass spectrometry was limited to analysis of elements and small chemical compounds. Together, these two technologies allow unknown proteins to be identified by comparing mass measurements of peptides derived from a protein to the theoretical peptides of all proteins encoded by an organism's genome.

MASS SPECTROMETERS

Mass spectrometers are very expensive and complicated instruments that do one specific thing: measure the mass (m) to charge (z) ratio of gas phase ions. Mass spectrometers are the world's most accurate and sensitive scales. They can measure the mass of individual molecules, sometimes to an accuracy of better than 0.001 daltons (remember that a typical protein might have a mass of 50,000 daltons). Mass spectrometers are so sensitive that only attomole quantities are needed for analysis in some instruments. (You may never have heard the term "attomole." An attomole is 10–^{18} moles.) Mass spectrometers are also incredibly fast, with current instruments capable of generating multiple spectra each second. Despite being technologically sophisticated and coming in a variety of types, all mass spectrometers share some common components (Figure 7.1). They all contain an ionization source whose job is to generate gas phase, charged molecules from samples. They all contain mass analyzers, which separate ions based on differences in their m/z values. And they all contain an ion detector connected to an external computer for generating the spectra. Finally, mass spectrometers operate under strong vacuum conditions; thus, ions can travel through them without bumping into neutral air molecules.

There are two basic types of ion sources and about a half dozen different types of mass analyzers that are commonly used to analyze proteins. The instrument used to analyze your peptides has a matrix-assisted laser desorption ionization (MALDI) source and a time-of-flight (TOF) mass analyzer. MALDI produces gas-phase ions by blasting a dried sample spot on a steel plate with an ultraviolet laser (Figure 7.2). The sample is first mixed

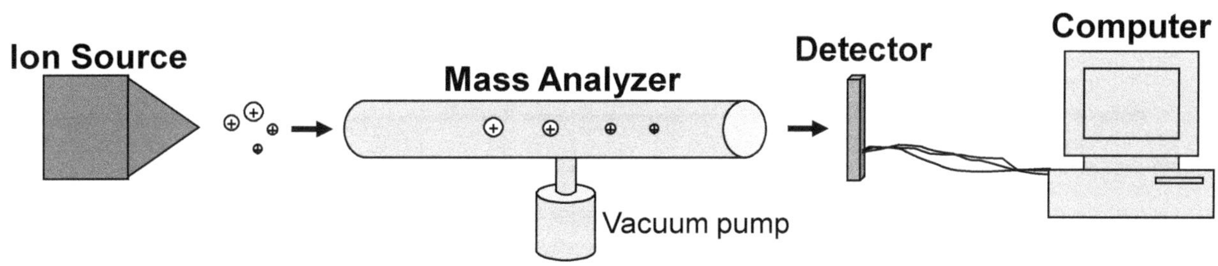

FIGURE 7.1. All mass spectrometers contain a common set of basic components. These include (1) an ionization source to generate charged, gas phase sample molecules; (2) a mass analyzer, under strong vacuum, to separate gas phase ions based on differences in mass/charge (m/z); (3) a detector; and (4) a computer to interpret and visualize ion signals at the detector.

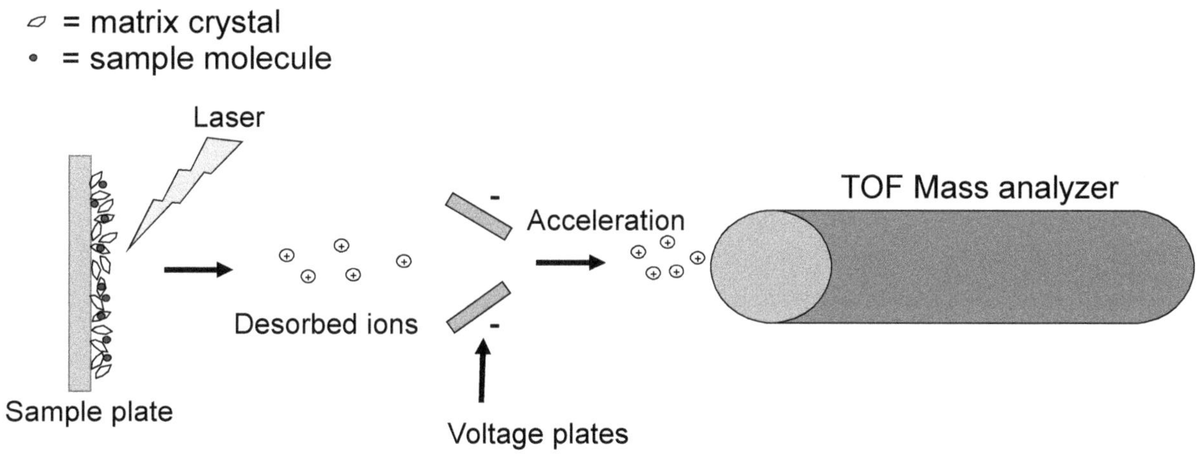

FIGURE 7.2. In matrix-assisted laser desorption ionization, crystals of a sample mixed with a light-absorbing matrix compound are irradiated with a laser pulse. The laser energy causes desorption of sample ions into the gas phase, where they are steered by electrical potentials into the mass analyzer.

with a chemical compound, called the matrix, that forms crystals with embedded sample molecules. The laser energy promotes transfer of protons from the matrix to the sample molecules, giving them the charge they need to be detected. The laser energy also promotes the desorption of sample molecules directly into the gas phase.

TOF mass analyzers are the simplest to understand (Figure 7.3). They consist of a vacuum tube with an ion entry port at one end and a detector at the other end. The detector simply measures the time needed for ions to travel the length of the tube. Ion velocity (v, distance per unit of time) is inversely proportional to m/z according to the following equation:

$$v = \sqrt{2Vz/m}$$

where m = mass, z = charge, and V = voltage used to accelerate ions.

This equation shows that ions with smaller m/z travel faster than those with larger m/z, with velocity inversely proportional to the square root of m/z. For ions of the same mass, higher charge results in higher velocity. Measured flight times are converted directly into m/z values using this relationship.

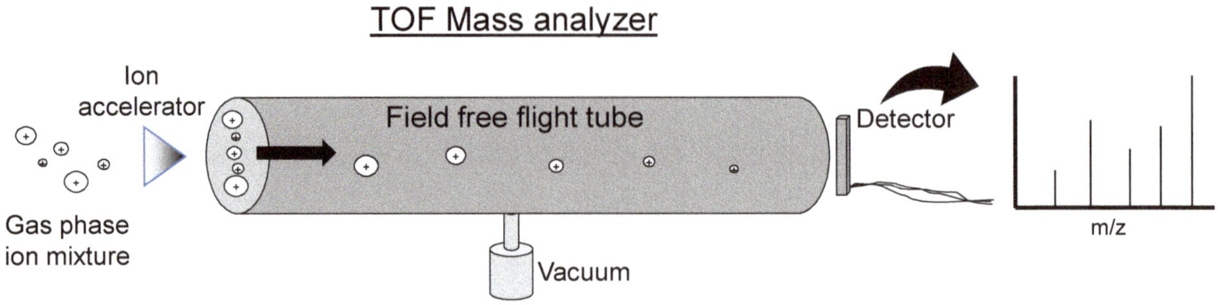

FIGURE 7.3. In a TOF mass analyzer, a mixture of ions is accelerated and injected into one end of a field-free tube under strong vacuum using an electrical voltage. Ions travel in the electrical field-free region at a velocity that is inversely proportional to the square root of mass/charge (m/z). A detector at the opposite end of the flight tube measures arrival of ions, and a computer converts detected signals into m/z and generates a composite spectrum of the mixture.

PROTEIN SAMPLE PREPARATION FOR MASS SPECTROMETRY

In today's lab you will be using mass spectrometry for one of its most common protein applications: identification of an unknown protein. Preparing samples for protein identification by mass spectrometry is remarkably easy. First, the proteins to be analyzed must be isolated in some form from the biological source. In your case, your protein has already been purified and run on SDS-PAGE, and the band has been visualized by Coomassie blue staining. The final step is to digest the protein with a site-specific protease such as trypsin. Protein bands of interest can be cut out of a polyacrylamide gel with a razor blade, placed in a microcentrifuge tube, and soaked in a trypsin solution. The trypsin enzymes will enter the gel pieces and digest the protein there. You performed this procedure last week. The resulting peptides are extracted by drying the gel slice with an organic solvent (this was done by the instructor or the teaching assistant the following day). The peptides are then ready for analysis in a mass spectrometer.

A site-specific protease will only cleave the polypeptide backbone of a protein at specific amino acids or amino acid sequences. Common examples that are useful for protein mass spectrometry analysis include trypsin and chymotrypsin (digestive enzymes). In contrast, proteases that cleave nonspecifically anywhere in a protein backbone, such as the digestive enzyme pepsin, are less useful for generating peptides for mass spectrometry analysis. This is because protein identification relies on a computer to calculate all possible peptides generated by protease digestion of the proteins encoded by a genome. There are too many possible peptides when the cleavage can occur at any position along the polypeptide backbone.

DATABASE SEARCHING TO IDENTIFY UNKNOWN PROTEINS FROM MASS SPECTROMETRY DATA

Computer algorithms have been created that can identify proteins based on the mass measurements of peptides generated from them. Using just peptide mass measurements to identify proteins is called peptide mass fingerprinting (PMF), and this is how it works (Figure 7.4). In the computer interface, the user provides the computer with the identity of the organism the protein sample is from (this is almost always known), the identity of a sequence database where that organism's protein sequences are stored, and the identity of the protease used to create peptides

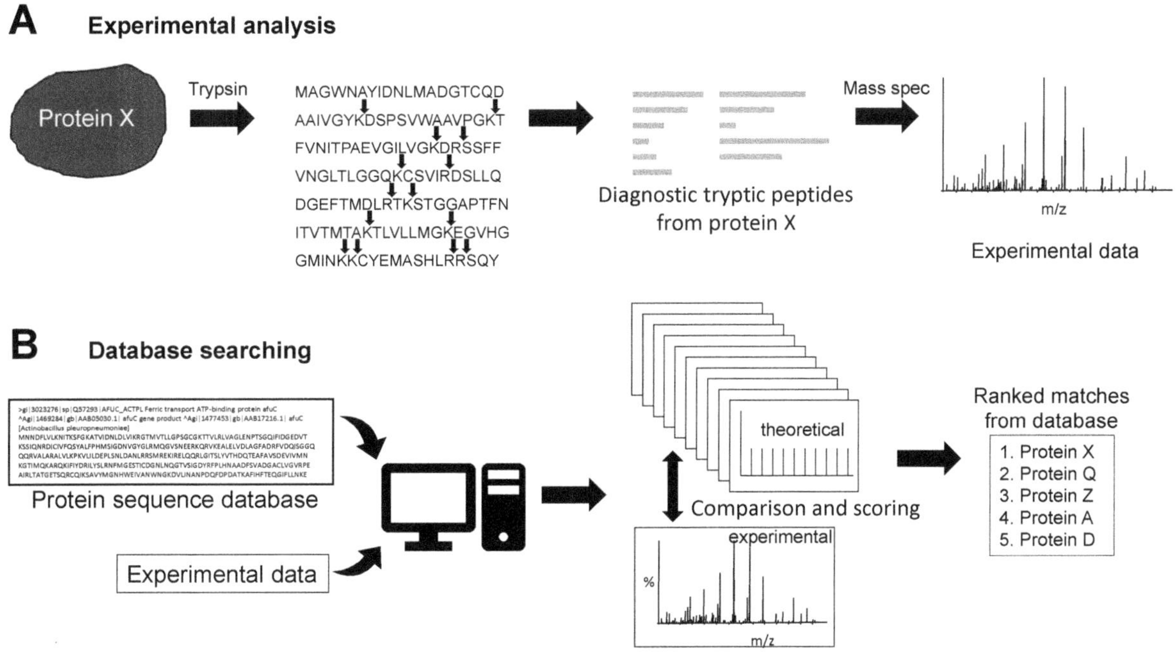

FIGURE 7.4. To identify proteins present in a sample by mass spectrometry, sample proteins are digested with a sequence-specific protease such as trypsin, and the resulting peptides are analyzed in a mass spectrometer. A computer generates theoretical spectra for all known protein sequences from the species used to generate the sample and compares each theoretical spectrum to the experimental spectrum. A score is assigned to each comparison based on number of matching peaks, and the computer returns a list of top-scoring sequence matches from the database with statistical probabilities that the matches could have occurred by chance.

(e.g., trypsin). The computer then takes the sequence of every protein encoded by the genome of that organism and creates a list of the theoretical peptides that would be generated by digestion with that protease. The computer compares the list of experimentally measured masses to the theoretical peptide masses from each protein in the database (with an error tolerance set by the user) and ranks the proteins from best to worst matches, assigning scores to each. Statistical analyses assess the probability that each match could have occurred by random chance. If the data are good, a single band excised from an SDS-PAGE gel should return one top protein match clearly above the other proteins. If more than one protein is present in the gel slice, multiple significant hits might be obtained.

THE FIELD OF PROTEOMICS

PMF is really a specialty experiment that doesn't take advantage of the true power of modern mass spectrometers. It is now possible to simultaneously identify and quantify thousands of proteins in complex biological samples. It is even possible to globally map posttranslational modifications on proteins using mass spectrometry. Such applications of mass spectrometry are the backbone of the growing field called proteomics. We don't have time to cover more of this broad topic in this course. But proteomic analyses using mass spectrometry rely on (1) effective separation of peptides by techniques such as high-performance liquid chromatography, (2) tandem mass spectrometry that provides sequence information on peptides in addition to their overall mass, and (3) extremely fast mass analyzers and sensitive detectors. Proteomics is having a huge impact on molecular life science research, drug development, and other areas of biotechnology.

OBJECTIVES

The objective of today's lab is to use a mass spectrum obtained from the trypsin digest of the putative FoCdc14 band on your Coomassie blue-stained gel last week to test if it is indeed the correct protein. The m/z peaks from the spectrum, reflecting the masses of trypsin-generated peptides from the protein band, will be used to search databases of *F. oxysporum* and *E. coli* proteins to look for sequences that could have generated the experimental spectrum. You expect that the only statistically significant match will be to the FoCdc14 protein. However, if you purified a contaminating protein from *E. coli*, then you expect to observe only a statistically significant match from the *E. coli* database. This exercise, which may be viewed as overkill in confirming that you have isolated the intended target protein, is a useful illustration of the power of mass spectrometry to identify unknown proteins in biological samples for proteomics applications.

LAB INSTRUCTIONS

PART 1. PEPTIDE MASS FINGERPRINTING DATABASE SEARCH

1. On one of your group's computers, open the mass spectrum image provided to your group. Also open Microsoft Excel. Create a peak list with the m/z values for all of the major peaks in your spectrum. You should copy the 20–30 most intense peak values observed. Save this peak list file to your group's data storage folder as well.
2. Now, open your internet browser and go to the Mascot database search engine website at https://www.matrixscience.com/. Click on the "Free Trial" option on the main menu bar and then "Access Mascot Server." On the next page click "Perform search" under the "Peptide Mass Fingerprint" section. This takes you to the search interface page, which should look similar to Figure 7.5.
3. The instructor will provide you with an email address to authorize server use. You can use your own, but you will need to go through a validation process.
4. Scroll down to the Data input section and select the "Query" radio button. Copy the values from the m/z peak list you made from your mass spectrum in step 1 above and paste them into the Query window.
5. Now it is time for the important part: parameter settings. Make sure trypsin is selected as your enzyme. Set missed cleavages at 1 for the initial search (you can try again with 0 missed cleavages later; ask your instructor for an explanation of this setting if you are not sure what it means). Leave the variable and fixed modification windows empty. Set peptide tolerance to 100 ppm. For the first search, set the database to "UP219602_F_oxysporum." Leave everything else as default. Your settings should look like those illustrated in Figure 7.5.
6. Click on "Start Search . . . ," and in a few seconds the results window should appear with your search results.
7. If you have purified the correct protein, you should have a single protein identified with a statistically significant score (a single red bar to the right of the green boxed area in the Mascot Score Histogram). Figure 7.6 shows an example of what your results page should look like and where you can find the important information on it. Save screenshots of your results page to your group's data storage folder. Make sure you capture the histogram, the identity and score of the top match, and the search parameters (located at the bottom of the page).

FIGURE 7.5. The Mascot database search algorithm's query submission page for analysis of peptide mass spectrometry data for protein identification. Users select the appropriate protein sequence database, taxonomy, the protease enzyme used for peptide generation, and other parameters such as missed cleavages allowed, peptide tolerance, and any peptide modifications that may exist. The list of measured m/z values from the experimental spectrum is entered in the query window or uploaded as a text file. The database search engine is available at www.matrixscience.com.

8. Click on the top hit in the match list below the histogram to go to a more detailed results page, where you will see information on the peptide sequences that were matched, the sequence coverage, and other useful information. Save screenshots of the peptide match information on this page as well. The instructor will provide more information on how to interpret the data on this page.
9. If you don't get any significant search scores, talk with the instructor about changes to the parameter settings that might improve your search and retry it. Changing the peptide tolerance setting is most likely to help. The alternative is that you have isolated a contaminating *E. coli* protein.
10. To confirm that the band is not a protein from the host organism, *E. coli*, repeat the search now with identical settings but select "UP625_E_coli_K12" as the database. You should expect no significant hits (no bars outside the green box in the histogram).

PART 2. BLAST SEARCH TO CONFIRM PROTEIN ID

Is the top hit from your *F. oxysporum* database search annotated as "Cdc14"? Probably not. Maybe it is just generically annotated as "protein tyrosine phosphatase." Since the genomes of many species are not annotated well, you need to use the sequence similarity search algorithm BLAST again to confirm that this is really the *F. oxysporum* CDC14 ortholog that you purified:

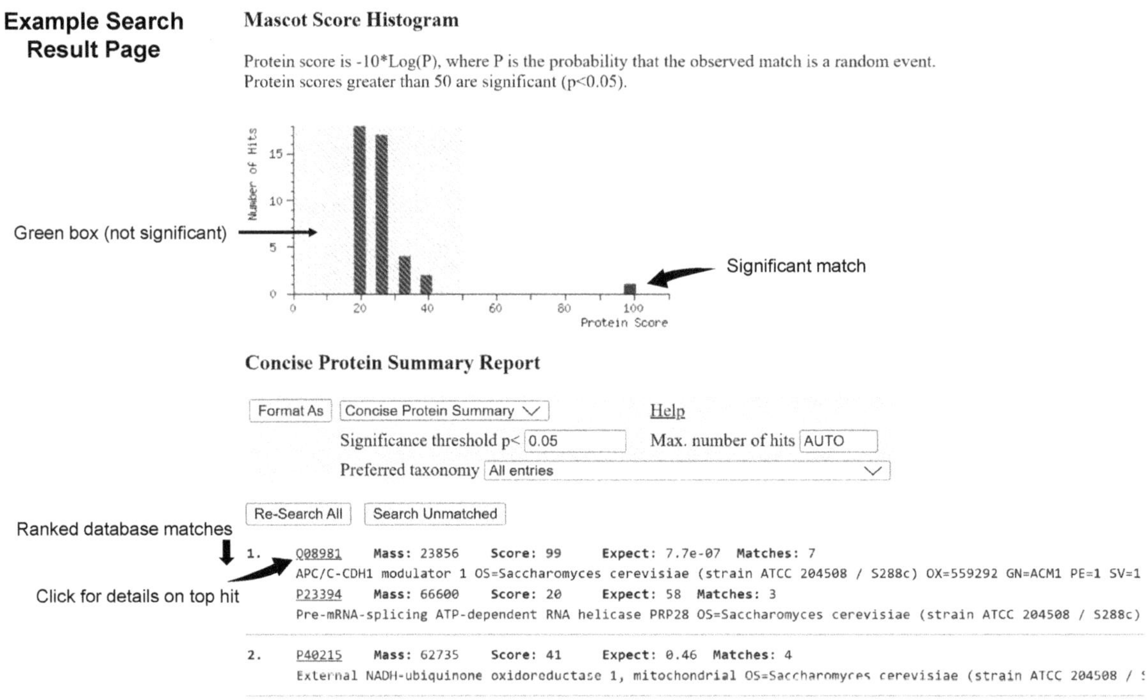

FIGURE 7.6. An example results page from a Mascot peptide mass fingerprint database search. The histogram provides a visual representation of top matches, with the right edge of the shaded region representing the 95% confidence boundary. Matches outside this region are statistically significant. In this example, a single statistically significant match was detected. The statistical definition of protein scores is provided, and links to additional data and information on the top matching proteins are found below the histogram. The database search engine is available at www.matrixscience.com.

1. This is easy to do. On the results page for your top hit (step 8 above) there will be a link for running BLAST at the NCBI site. Just click on it.
2. On the Protein BLAST search page, all you need to do is select "ClusteredNR" as the Database and "*Saccharomyces cerevisiae*" as the Organism. Click the BLAST button at bottom to run the search.
3. Save the results of this search as a screenshot showing the name of the top match. This should reveal conclusively if you have the correct protein purified.

DATA ANALYSIS AND SUBMISSION

Save a Word file containing the following information to your group data storage page and upload it to the course learning management system before you leave today:

1. Results of your Mascot database search. Provide (1) the peak list you used for the search; (2) screenshots of the main search results page for both searches showing the histogram, the identity and protein score of the top hit, and the search parameters; and (3) screenshots of the detailed results page for any significant matches, including the list of the peptides that were matched along with the protein sequence coverage map.
2. BLAST search results addressing whether or not the identified protein is a Cdc14 ortholog.

8. PROTEIN HYDRODYNAMICS

BACKGROUND AND THEORY

Many proteins function as components of multisubunit complexes. Complexes of multiple proteins are called oligomers. A two-protein complex is a dimer, three proteins are a trimer, four proteins are a tetramer, etc. Complexes can either be homo-oligomeric, meaning they consist of multiple copies of the same protein, or can be hetero-oligomeric, meaning the subunits are different proteins. Oligomerization is often referred to as the "quaternary structure" of proteins. Restriction endonucleases (you learned about those in the molecular cloning lecture in Week 3) are classic examples of homo-oligomeric proteins, consisting of two identical catalytic subunits, one to cut each DNA strand (Figure 8.1). The histone core of nucleosomes (Figure 8.2) is a hetero-oligomeric complex of four different histones, each present in two copies, that compacts DNA into chromatin and regulates gene expression and other DNA processes. One aspect of characterizing and understanding how proteins work is to determine their functional mass, or oligomeric structure. If a protein acts in complex with other proteins, then its functional mass might be very different than that suggested just by its amino acid sequence or its migration position during SDS-PAGE. One way to determine the oligomeric state and mass of proteins is to measure their *hydrodynamic properties*. Hydrodynamics refers to the motion of

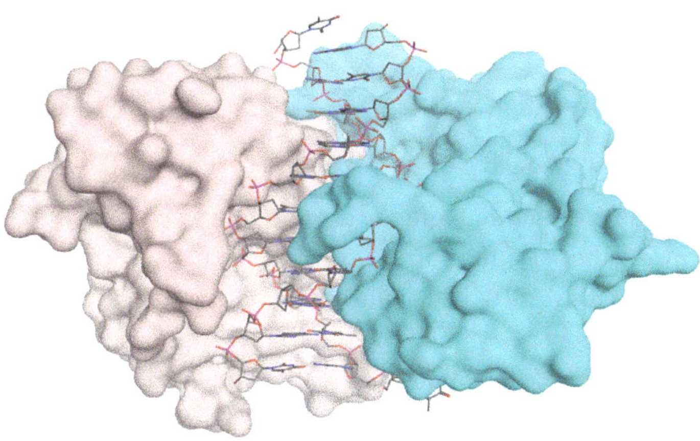

FIGURE 8.1. Structure of the restriction endonuclease, BamHI, bound to its target DNA sequence. Like most restriction endonucleases, BamHI is a homodimer (the two subunits are colored uniquely in surface representation), with each subunit catalyzing cleavage of the identical site within the recognition sequence on opposite DNA strands. The DNA is depicted in stick format. Image created from RCSB protein databank ID 1BHM and generated with Molecular Operating Environment.

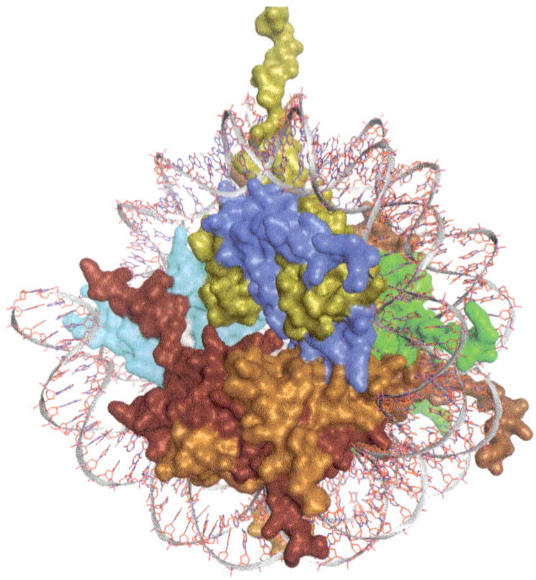

FIGURE 8.2. The core histone octamer of eukaryotic nucleosomes is an example of a heteroligomeric protein complex. The octamer consists of four homodimeric proteins: H2A, H2B, H3, and H4. The eight subunits, depicted in surface mode, are each colored uniquely, and the DNA that wraps around the histone core is shown in stick mode. Image created from RCSB protein databank ID 1AOI and generated with Molecular Operating Environment.

particles (such as proteins) in solution. Using various experimental techniques, measuring the solution motion of particles can provide useful information about their size, shape, and density.

In this week's lab, you will measure a hydrodynamic property of your purified protein called the Stokes radius by size exclusion chromatography (SEC). The Stokes radius provides a good estimate of molecular mass for globular proteins. We will then combine this measurement with another hydrodynamic value, the sedimentation coefficient (data provided to you), to calculate a more accurate molecular mass. As you will see, these two properties used together can determine reasonably accurate masses of any proteins and protein complexes in solution regardless of their shape.

SIZE EXCLUSION CHROMATOGRAPHY AND STOKES RADIUS

SEC, sometimes called gel exclusion chromatography or gel filtration chromatography, is a common method used to separate proteins by size but also to estimate their molecular mass. The principle on which SEC is based is illustrated in Figure 8.3. Basically, SEC columns contain porous beads made of a material such as cross-linked acrylamide or polysaccharide. Small molecules can enter the pores and get "stuck" or retarded as they flow through the column bed. Large molecules will mostly flow around the beads because they can't enter the pores and will therefore migrate through the column faster. So, the larger a molecule is, the faster it flows through the column.

The Stokes radius represents the radius of a sphere that would have the same diffusional properties in solution as the protein being studied (Figure 8.4). If all proteins were perfectly spherical, then determining Stokes radius alone would be an accurate way to measure mass. However, Stokes radius is affected by protein shape,

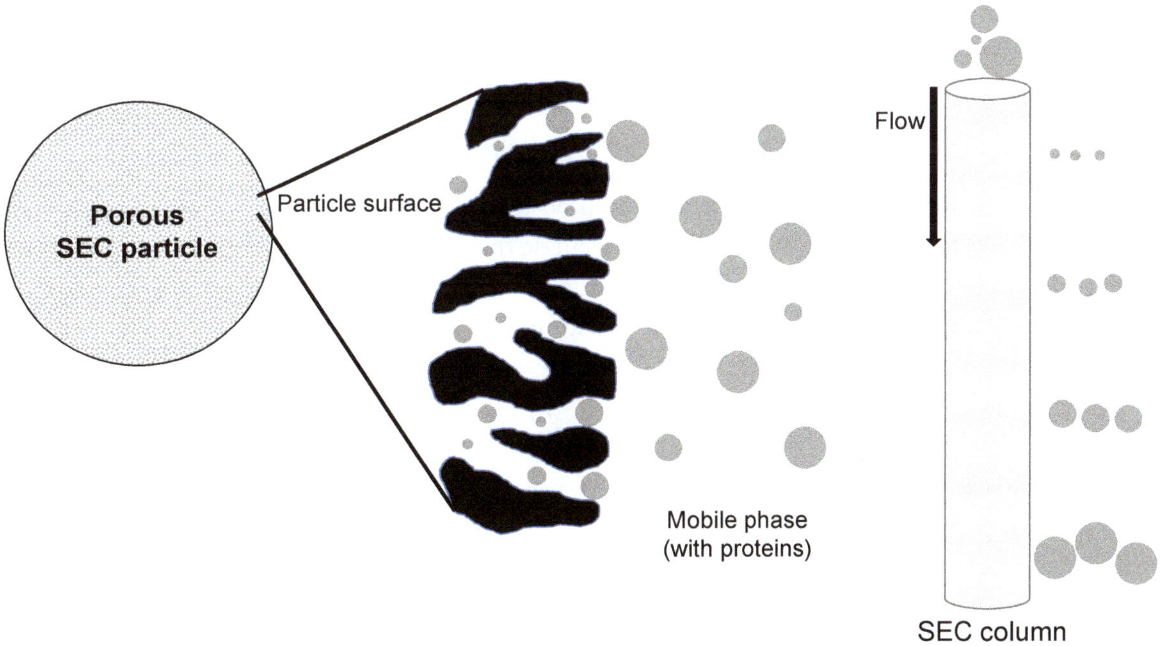

FIGURE 8.3. In SEC, the stationary phase particles (left) are porous with channels of varying width that are differentially accessible to mobile phase proteins, depending on their size and shape. During migration through the column, large proteins are sterically excluded from the particle channels, causing them to exit the column fastest. Small proteins frequently enter the particle channels, extending their migration time and causing them to exit the column more slowly. Protein mixtures are effectively separated into size fractions based on protein size and shape, as indicated in the diagram on the right.

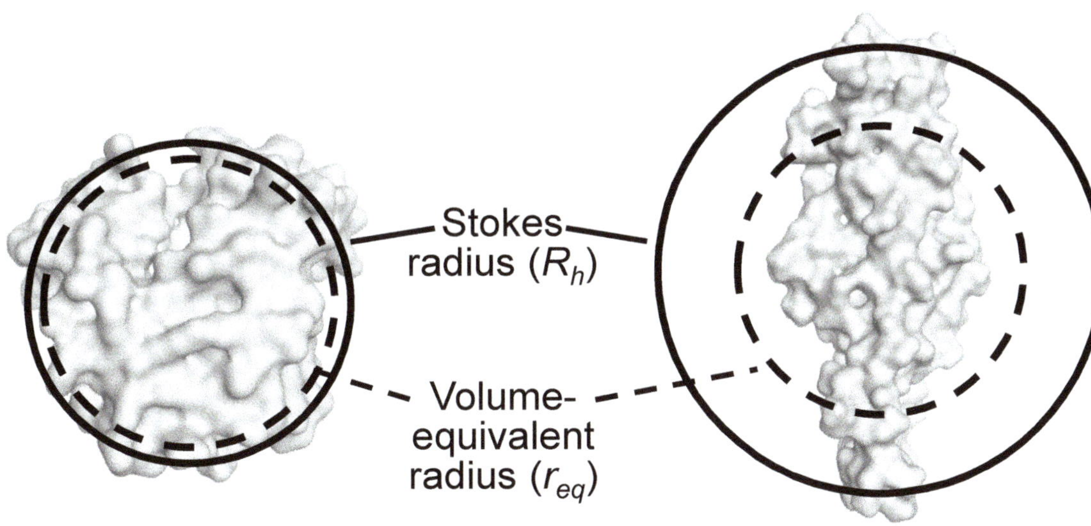

FIGURE 8.4. The hydrodynamic radius of a protein (R_h, solid circles), also called the Stokes radius, is the hypothetical sphere that a protein behaves like in solution. It is affected by protein shape as shown here, where two proteins of identical mass can have very different Stokes radii. A spherical protein has a much smaller Stokes radius than a highly elongated protein because the 3D rotation of an elongated protein causes it to behave as a larger sphere. For comparison, the identical volume-equivalent spherical radius (r_{eq}) of each protein, defined as the sphere of identical volume, is shown (dashed circles).

and most proteins are not perfect spheres. Thus, measuring Stokes radius by SEC does not provide an accurate mass measurement for proteins that do not have a globular or spherical shape.

ANALYSIS OF PROTEIN HYDRODYNAMICS BY CENTRIFUGATION

Analytical ultracentrifugation is a powerful method used to directly measure the rate at which proteins (or particles) sediment during centrifugation and then calculate their mass (and oligomeric state) from this value. The rate at which a particle moves during centrifugation is known as its sedimentation coefficient (s, also reported in Svedberg units, S) and is defined by the equation

$$\frac{v}{\omega^2 r} = \frac{m_0(1 - v\rho)}{f}$$

where $v/\omega^2 r$ is the sedimentation coefficient; $m_0(1\text{-}v\rho)$ is the effective mass (v and ρ are the partial specific volume, or buoyancy, of the particle and the density of the solution, respectively); and f is a frictional coefficient. A Svedberg is equal to 10^{-13} seconds, and thus sedimentation coefficient has units of time.

You will find in the lecture that sedimentation of a particle in a centrifuge is dependent on two opposing forces: the centrifugal force and the frictional force (Figure 8.5). Centrifugal force depends on a particle's mass, the speed of the centrifuge, and the distance of the particle from the axis of rotation. This is why apparent mass can be measured using centrifugation. The frictional force is dependent on the viscosity of the solution and the size and shape of the particle.

Analytical ultracentrifugation requires sophisticated expensive instrumentation and expertise in using it and analyzing the data. Rate zonal density gradient centrifugation is another simpler method commonly used

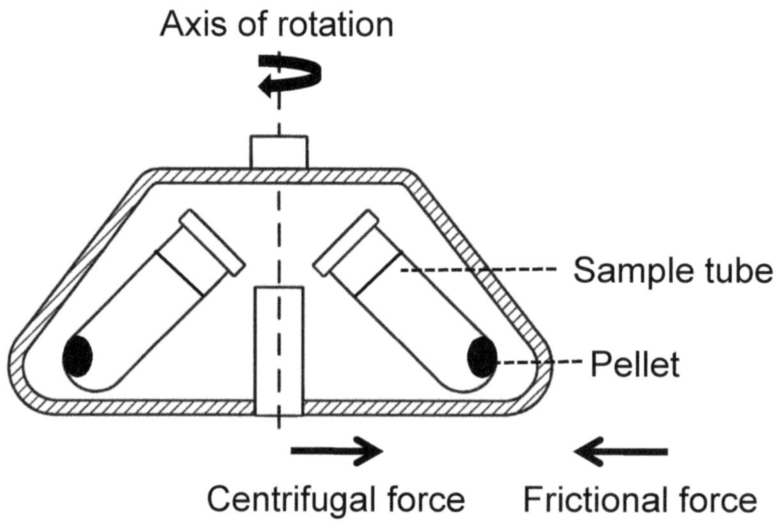

FIGURE 8.5. The principle of centrifugation. When liquid samples are subjected to centrifugation, particles such as proteins experience two opposing forces. The spinning motion generates centrifugal force that pushes particles outward, perpendicular to the axis of rotation. The centrifugal force depends on particle mass and distance from the axis and on the speed of rotation. Particles also feel an opposing frictional force as they sediment through the sample solution. The frictional force depends on the solution viscosity and the size and shape of the particles.

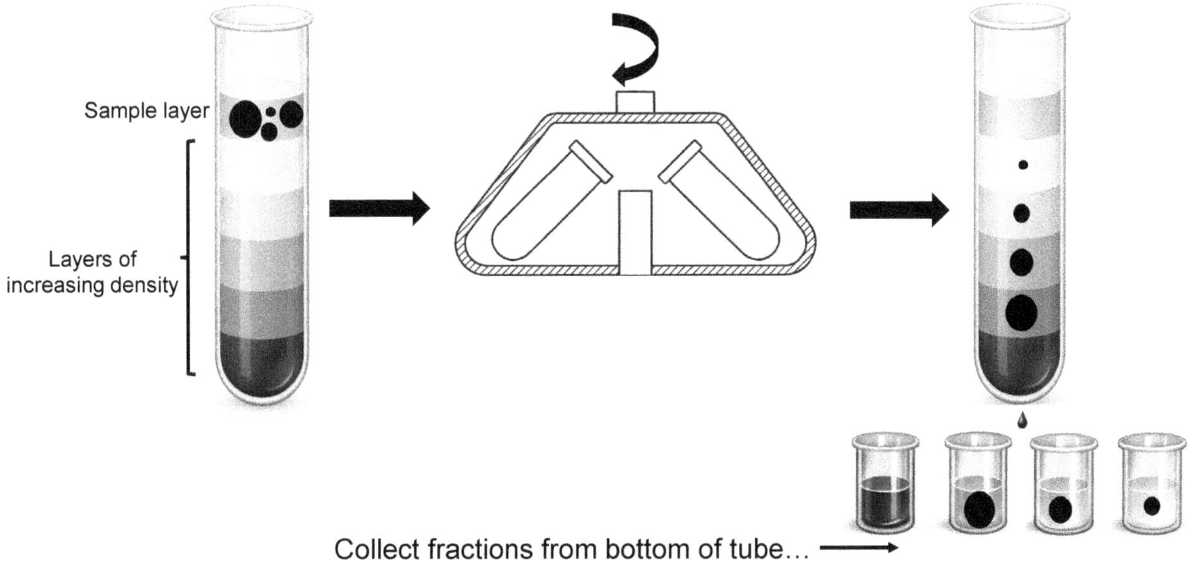

FIGURE 8.6. Rate zonal density gradient centrifugation involves the slow sedimentation of protein samples through solution layers of increasing density. Penetration of proteins through the density gradient is dependent on their mass and shape. After centrifugation is complete and sufficient separation is achieved, fractions are collected from the bottom of the tube and analyzed by SDS-PAGE or some other detection method. Inclusion of standard globular proteins of known mass and sedimentation coefficient allows generation of a standard curve and estimation of the mass and oligomeric state of experimental proteins.

to estimate protein molecular weight and requires only a standard ultracentrifuge. In a rate zonal experiment, particles sediment in a gradient of increasing density (usually made from sucrose or glycerol solutions). The dense solvent helps establish defined stable protein bands separated by differences in sedimentation rate (Figure 8.6). Again, the sedimentation rate is directly related to the protein's mass and shape. This method does not require the direct visual measurement of sedimentation velocities required with analytical ultracentrifugation. Instead, a set of spherical protein standards with previously defined S values is used to establish a standard curve relating S to migration position in the gradient. Since the standards are all globular, eliminating shape effects, the curve can also be interpreted as mass versus migration position and can therefore be used to determine apparent mass of experimental proteins. Remember, though, that if a protein being analyzed is not spherical (most proteins are not), the measured mass value will be inaccurate.

USING STOKES RADIUS AND SEDIMENTATION COEFFICIENT TO CALCULATE MOLECULAR MASS

Hopefully from the paragraphs above it is clear that both SEC and density gradient centrifugation can be used to estimate the mass of proteins in solution but that the mass values will be inaccurate for proteins with nonspherical shapes. The more a protein deviates from a sphere, the more inaccurate the mass estimate using these approaches. In 1966, Siegel and Monty discovered and reported that the true molecular mass of a protein could be determined by combining the measured Stokes radius and sedimentation coefficient.[1] The shape effects encountered in SEC and density gradient centrifugation cancel out when the following equation is used to determine mass:

$$MW = (6\pi\eta Nas)/(1-\nu\rho)$$

where η is the solution viscosity in g/cm sec (.01005 for water), **N** is Avogadro's number (6.022×10^{23}), **a** is the measured Stokes radius in cm, **s** is the measured sedimentation coefficient x 10^{-13} sec, ν is the average partial specific volume of the protein in cm^3/g, and ρ is the solution density (usually 1.0 g/cm^3 is used).

OBJECTIVES

This week your goal is to measure the functional molecular weight and oligomeric state of your FoCdc14 protein using the method of Siegel and Monty. You will generate your own Stokes radius value and estimation of MW by SEC alone. You will be provided with density gradient centrifugation data to analyze and determine sedimentation coefficient and also estimate MW (density gradient centrifugation takes too long to be practical in a lab course). You will then use the Stokes radius and sedimentation coefficient that you measured, along with the Siegel and Monty equation, to calculate the true shape-independent MW and make a conclusion about the oligomeric state of FoCdc14. You will compare that to what is known about the oligomeric state of other Cdc14 proteins.

LAB INSTRUCTIONS

PART 1. SIZE EXCLUSION CHROMATOGRAPHY ANALYSIS

Overview: Each group will inject mixtures of protein standards (with known Stokes radii) onto a size exclusion column (particles made of a synthetic polymer called Sephacryl S-200) using the same ÄKTA Start systems used previously for protein purification. Protein elution will be monitored by UV light absorbance (remember, proteins absorb strongly at 280 nm mostly due to aromatic amino acids such as tryptophan). The protein elution peaks will be used to create a standard curve of elution volume versus Stokes radius or molecular mass. Then each group will run their pure FoCdc14 protein samples and determine elution volume by UV absorbance. Stokes radius and apparent MW will be calculated using the standard curve.

In the morning lab session, inject 1 ml of the standard protein mix for standard curve generation and run the method BCHM322_SEC.

1. Make sure you have your flash drive inserted in the ÄKTA Start side USB port.
2. Fill the provided 12 ml plastic syringe with SEC column buffer, attach it to the front port on the injection valve, and push a few ml of buffer through the green coiled sample loop. *Do not remove the syringe at this point.*
3. Use a 1 ml syringe to draw up as much of the standard mix as you can. Carefully push all air out of the syringe so you don't introduce air into the column.
4. Unscrew the waste line from the bottom of the injection valve and replace it with the provided plug. Now you can remove the syringe and replace it with the 1 ml syringe containing your standard mix, but *don't push the mixture into the sample loop yet.*
5. Remove the plug and screw the waste line back in to the bottom of the injection valve.
6. Start the SEC method BCHM322_SEC from the System Control window (Home → Method Run → User defined). You will need to provide a two-digit file name number again (01 is usually fine). When prompted to switch

the injection valve from Load to Inject: 1) push the standard mixture into the sample loop by depressing the syringe plunger completely. 2) move the lever on the top of the injection valve from the Load (left) to the Inject to Column (right) position. 3) press "Continue" on the touch screen. You can now remove the syringe as well. You will be prompted in a minute to move the lever back to the Load position and press "Continue" again. The method will run to completion, and the file will be saved for analysis in the afternoon lab period.

In the afternoon lab session, run pure FoCdc14 on the column and analyze the data.

1. Dilute 250 µl of your pure FoCdc14 protein 5-fold in SEC column buffer. Mix well and then spin for five minutes in a microcentrifuge at max speed to pellet any insoluble material. Draw the sample up into a 1 ml syringe and displace any bubbles.
2. Repeat the steps from the morning session to run your pure protein through the column using the same BCHM322_SEC method (now using "02" as the file name).
3. Once your protein peak has eluted, the run can be stopped. Wait for the file to be saved before removing the flash drive from the USB port.
4. Open the standards and FoCdc14 .bmp files on your computer and find the elution peaks.
5. Estimate the elution volumes for each peak (point of maximum absorbance). Record the peak elution volumes (V_e) for each standard protein and FoCdc14.
6. In your group's data storage folder for this week you will find another .bmp file called "group#_bluedextran.bmp." Open that file and record the peak volume observed in the chromatogram. Blue dextran is a large polysaccharide that is completely excluded from the SEC resin and defines the void volume (V_0), which represents the earliest possible elution position for the column.

Create two standard curves in Excel. First, plot $\text{Log}_{10}(\text{MW})$ using the values found in Table 8.1 on the Y-axis as a function of elution volume. The plot should be linear with a negative slope. Second, plot Stokes radius values from Table 8.1 as a function of the square root of $(-\log_{10}(K_{av}))$,

$$K_{av} = (V_e - V_0)/(V_t - V_0)$$

where V_t is the total column bed volume (120 ml for the Sephacryl S-200 HR columns), V_e is the peak elution volume of each standard, and V_0 is the column void volume. This plot should be linear with a positive slope. Perform linear regression for each curve to generate the standard curve equation.

Finally, using the experimentally measured V_e value for your pure FoCdc14 protein, calculate both its Stokes radius (in nm) and the apparent mass (in Da).

TABLE 8.1. *SEC Standards Information*

STANDARD PROTEIN	MASS (Da)	STOKES RADIUS (nm)
Ribonuclease A	13,700	1.64
Ovalbumin	44,000	3.05
Alcohol dehydrogenase	150,000	4.55

PART 2. ANALYSIS OF SUCROSE DENSITY GRADIENT CENTRIFUGATION DATA

Experiment description. Final density gradients were separated into fractions using a peristaltic pump starting at the bottom of the gradient (largest proteins). Equal samples from each fraction were run on a gel. The gel was stained with Coomassie blue and imaged on a Bio-Rad digital imaging system. The fraction with the highest intensity value (peak) for each protein was determined using Bio-Rad Image Lab software. The resulting data for the standard proteins are shown in Table 8.2. When FoCdc14 was analyzed by density gradient centrifugation, its peak intensity was detected in fraction 18.

Make two standard curves again. The first should plot sedimentation coefficient as a function of peak fraction number. The second should plot mass as a function of peak fraction number. Both should be linear, with a negative slope. Perform linear regression to get the equation and use it to calculate both apparent mass and sedimentation coefficient for FoCdc14.

PART 3. CALCULATE MASS AND OLIGOMERIC STATE BY THE METHOD OF SIEGEL AND MONTY

You have calculated the Stokes radius and the sedimentation coefficient for FoCdc14.

Now, use the equation of Siegel and Monty to calculate the shape-independent molecular mass of FoCdc14,

$$MW = (6\pi\eta N a s)/(1-\nu\rho)$$

where

a is the Stokes radius (needs to be provided in cm—your curve will be plotted in nm).
s is the sedimentation coefficient (units are x 10^{-13} sec).
η is the viscosity of the solution. Use the value for water (0.01005 g/cm sec).
N is Avogadro's number (6.022 x 10^{23} /mole).
ν is the partial specific volume of the protein, which can be calculated from amino acid composition but generally doesn't change much from one protein to another. Use the average protein value of 0.724 cm^3/g.
ρ is the density of the solution. Use the value for water (1.0 g/cm^3).

TABLE 8.2. *Density Gradient Standards Information*

STANDARD PROTEIN	MASS (Da)	SEDIMENTATION COEFFICIENT (S)	PEAK FRACTION
Serum albumin	66,000	4.3	20
Aldolase	160,000	7.35	14.5
Catalase	240,000	11.3	9

DATA ANALYSIS AND SUBMISSION

Make sure the .bmp files of your two SEC runs are saved to your data storage folder. In addition, create and save a single Excel file containing the following items and upload it to the course learning management system before leaving lab today:

1. Standard curves—you should have two from the SEC standards and two from the density gradient standards (four total)—properly labeled with correct units. Create separate sheets labeled "SEC" and "Density" for these.
2. Molecular weight calculation of FoCdc14 from SEC data alone using your MW versus V_e standard curve. Show work in the SEC sheet using Excel functions.
3. Molecular weight calculation of FoCdc14 from the density gradient data using your mass versus peak fraction standard curve. Show work in the Density sheet using Excel functions.
4. Molecular weight calculation of your pure protein using the method of Siegel and Monty. Show all work in Excel on a separate sheet named "Siegel-Monty."
5. Conclusion about the oligomeric state of your protein, included on the Siegel-Monty sheet.

9. ASSAYING ENZYME ACTIVITY

BACKGROUND AND THEORY

CHEMICAL REACTION RATES AND CATALYSIS

The rates of chemical reactions are dependent on the concentrations of the reactants. The term "rate" refers to the change in reactant or product concentrations over time. A first-order reaction is dependent on the concentration of a single reactant, and a second-order reaction is dependent on the concentration of two reactants. An equation describing the rate of any chemical reaction can be written based on this simple rule, as illustrated in Figure 9.1. In these equations, each step has a unique rate constant, k, determined by the physical and chemical factors that define how easily that step occurs.

Because the concentrations of the reactants are constantly decreasing as the reaction progresses, the rate is also constantly changing, as shown by the example reaction progress curve in Figure 9.1. Chemical reactions are usually reversible, and the same rate principles apply to the forward and reverse reactions. Thus, the overall rate of a reaction at any given point in time is equal to the rate of the forward reaction minus the rate of the reverse reaction. Equilibrium is defined as the condition when the rates of the forward and reverse reactions are equal and there is no net change in reactant or product concentrations. All reactions naturally proceed toward the point of equilibrium.

Catalysts are substances that accelerate the rates of chemical reactions without changing fundamental properties of the reaction or being altered themselves. They alter the rate constants of reaction steps but do not alter the overall reaction equilibrium (Figure 9.2). Catalysts change rate constants by lowering the activation energy needed for the conversion of reactants to products. Catalysts are incredibly important in biological systems.

ENZYMES

Biological molecules that catalyze chemical reactions are called enzymes. Most enzymes in biological systems are proteins; however, there are exceptions to this, including some catalytic RNA molecules. Reactants in enzyme-catalyzed reactions are called substrates. A unique feature of enzymes, compared to chemical catalysts, is that enzymes bind their substrates in specific orientations at dedicated active sites through molecular interactions and shape complementarity, which will be discussed in Chapter 11. Thus, the formation of a physical enzyme-substrate (ES) complex becomes the first step in enzyme-catalyzed chemical reactions (Figure 9.3). The binding of enzyme to substrate is reversible, and substrate can dissociate from enzyme without being converted to product.

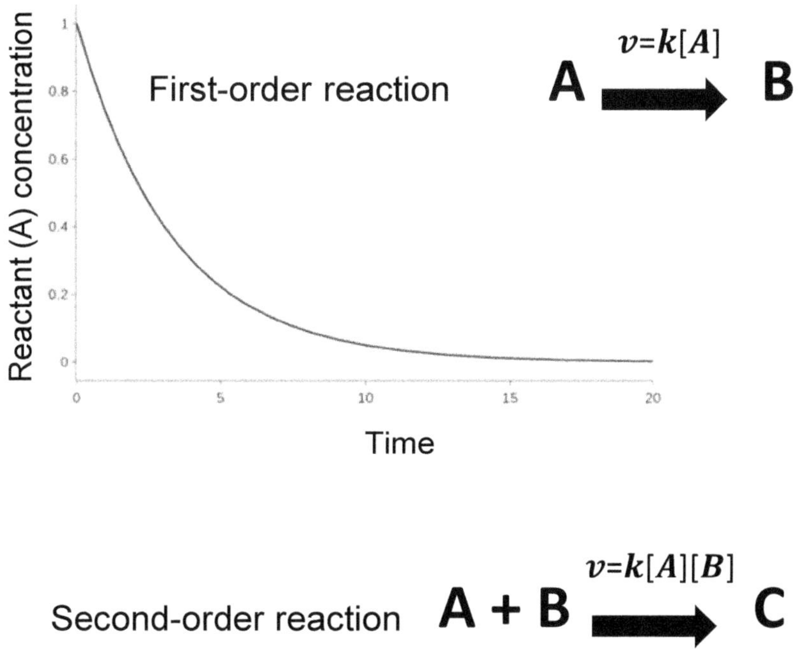

FIGURE 9.1. Reaction progress curve for a first-order chemical reaction. The concentration of reactant (A) decreases over time as it is converted to product (B). As A decreases, the reaction velocity (v) slows according to the indicated rate equation. This behavior is characteristic of unimolecular reactions and many biochemical processes such as enzyme-catalyzed reactions under certain conditions. The rate equation for a second-order reaction dependent on the concentrations of two reactants is shown under the first-order progress graph.

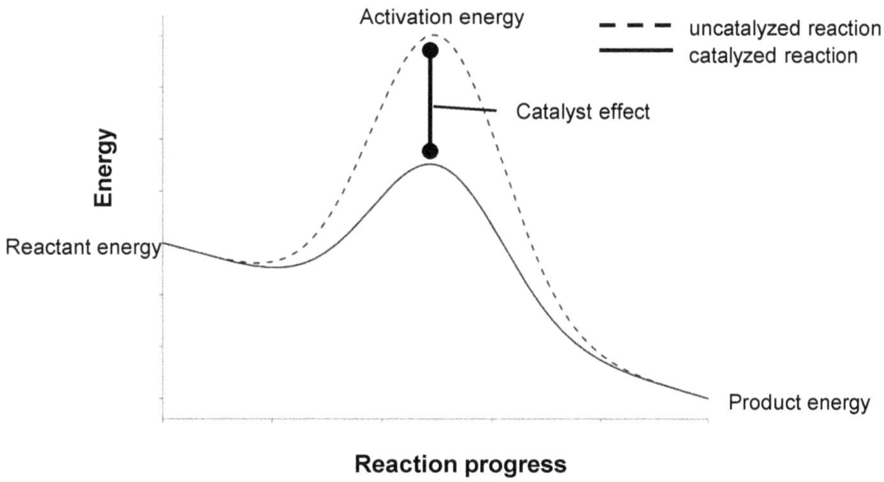

FIGURE 9.2. The energy progress curve of a spontaneous chemical reaction (products have lower free energy than reactants) is shown with and without a catalyst present. An activation energy barrier must be surpassed for the reaction to proceed. Catalysts accelerate reactions by reducing the activation energy required. They do not alter the free energy of the reactants and products or the reaction equilibrium.

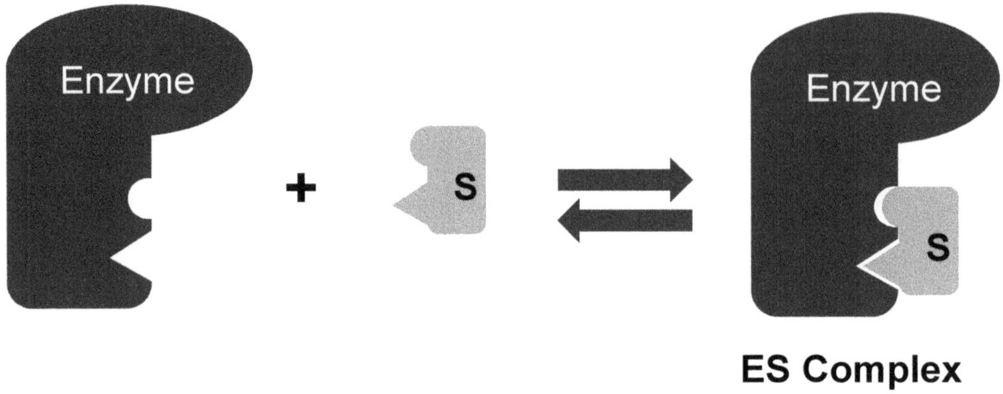

FIGURE 9.3. One key difference between an enzyme (E) and conventional chemical catalysts is that enzymes form physical complexes with substrate molecules (S) prior to the chemical transformation. Substrates bind at defined sites on the enzyme in specific configurations that favor the chemical reaction.

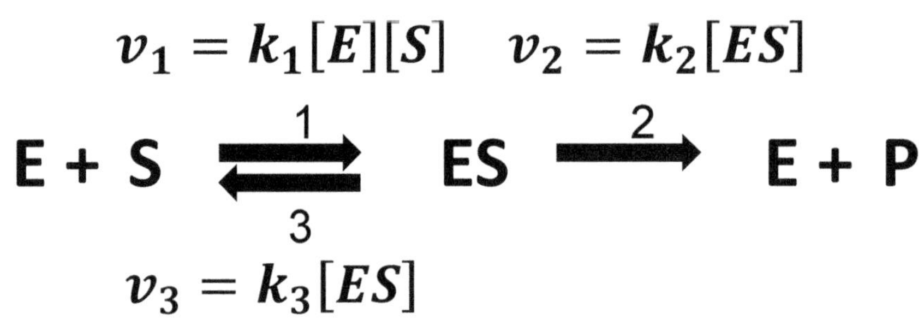

FIGURE 9.4. The simplest possible enzyme reaction scheme is shown, where enzyme (E) and a substrate (S) form a complex (ES) followed by conversion of S to product (P) and release. Binding of E and S is reversible, with each direction described by a unique rate equation. In the lab, where reactions are initiated in the presence of S and the absence of P, the reverse reaction is negligible and is not considered.

Rate equations for enzyme-catalyzed reactions follow the same rules as those for uncatalyzed chemical reactions described above. The simplest possible enzyme reaction scheme, which includes formation of the ES complex, conversion of substrate to product (P), and release of the product to regenerate the free enzyme is illustrated in Figure 9.4, along with its associated rate equations. Note that many enzymes catalyze more complicated reactions with additional steps, sometimes with multiple substrates and/or multiple products.

Enzymes are responsible for most of the metabolic processes occurring in cells. Enzymes have several amazing properties compared to chemical catalysts, which collectively allow the complex lifeforms on Earth to exist:

- Enzymes can catalyze chemical reactions at neutral pH, ambient temperature, and atmospheric pressure. Most chemical catalysts require one or more extreme conditions not suitable for life.
- Rate enhancements for enzyme-catalyzed reactions are far greater than those of chemical catalysts. Some enzymes can accelerate reactions trillions to quadrillions of times faster than the uncatalyzed reactions.

- Enzymes exhibit exquisite specificity for substrate structures, owing to the nature of their active sites that bind substrate molecules. For example, DNA and RNA polymerases catalyze very similar reactions, but one uses only deoxynucleotides as building blocks, and the other one uses only nucleotides, complex substrate molecules that differ by a single oxygen atom on the sugar portion.
- Enzymes can be dynamically regulated. Their catalytic activity can be turned on and off in response to environmental conditions to ensure that they act only when needed. Chemical catalysts cannot be dynamically regulated.

MEASURING ENZYME ACTIVITY WITH ASSAYS

Enzyme-catalyzed reactions are studied in the lab with experiments called assays. An enzyme assay is simply an experiment designed to measure the rate at which one or more substrate molecules are converted to one or more product molecules. An assay requires defined conditions with known concentrations of enzyme and substrate molecules and a method and analytical instrument for measuring either the consumption of substrate or the accumulation of product over time. Measuring the accumulation of product is by far the most common way to monitor reaction progress. Spectroscopic techniques such as absorbance and fluorescence are frequently used for product measurement.

There are several fundamental assay properties that determine their utility in measuring enzyme-catalyzed reactions (Figure 9.5). Linear dynamic range refers to the range of values where the instrument signal is proportional to product (or substrate) concentration. Limit of detection refers to the lowest product (or substrate) concentration that can be detected on the assay instrument. Finally, specificity of the measured signal

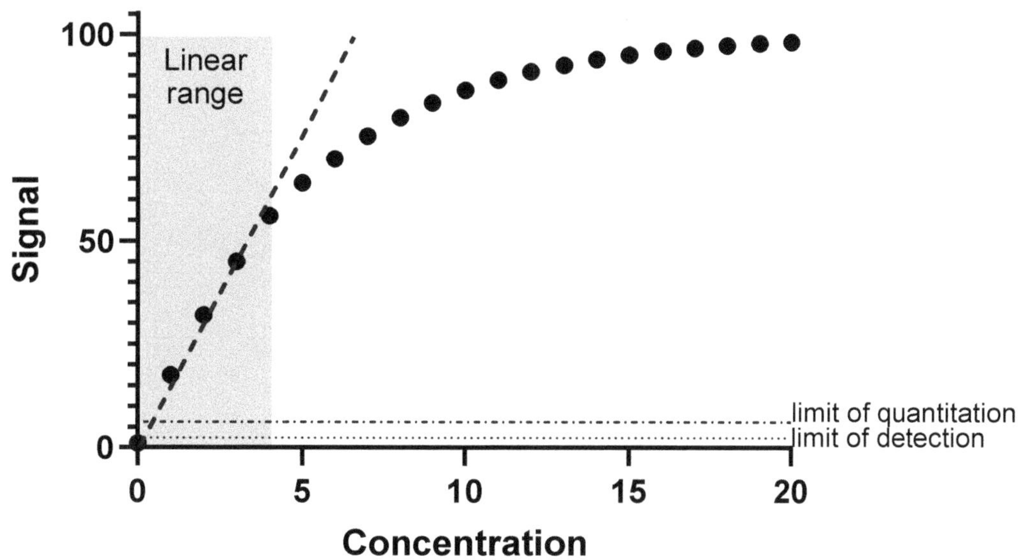

FIGURE 9.5. Several fundamental properties of enzyme assays that must be considered during design or selection are illustrated in this graph. All assays have a finite range where measured signal (e.g., absorbance) is directly proportional to product (or substrate) concentration (shaded area here). This is called the linear dynamic range. Above this range, signal response becomes nonlinear due to, for example, instrument limitations. At the lower end of the linear dynamic range the signal approaches the background, or noise level. The limit of detection is the lowest signal that can be reliably distinguished from noise, often defined as three standard deviations above background. The limit of quantitation is the lowest signal that can be reliably measured accurately and is often defined as 10 standard deviations above background. These values illustrate the sensitivity of an assay to measure the intended molecule(s).

is critically important. Interference by or contributions of other molecules in the reaction to the measured signal will confound the results. All of these properties must be considered when choosing or designing an assay to study a particular enzyme.

THE STEADY-STATE CONDITION

The steady state is a special equilibrium condition for an enzyme-catalyzed reaction where the concentration of the ES complex remains constant over a significant period of time in which rate measurements are being made. Importantly, the observed (or initial) velocity of the reaction is constant under steady-state conditions, unlike the changing reaction rate for standard chemical reactions shown in Figure 9.1. The steady state is a convenient and very common condition for scientists to use in studying enzymes and enzyme inhibitors because they do not need to consider effects of time on observed reaction rates. This makes experimental design much easier and the mathematical interpretation of data much simpler. The steady state is also physiologically relevant, as many enzymes operate under approximately steady-state conditions in biological systems.

How is the steady state established in experimental enzyme assays? The enzyme concentration must be very low relative to the substrate concentration, low enough that the total amount of substrate consumed while the reaction is being monitored is minimal (no more than 10% is a general rule). Since the formation of ES complex is dependent on the substrate and enzyme concentrations, limiting the consumption of substrate during the reaction ensures that the substrate concentration doesn't change much, and therefore the rate of ES formation doesn't change significantly. Also, since reactions begin with zero product under lab conditions, there is no significant reverse reaction for most enzymes under these conditions. The steady-state condition will result in a linear reaction progress curve (product concentration vs. time) (Figure 9.6).

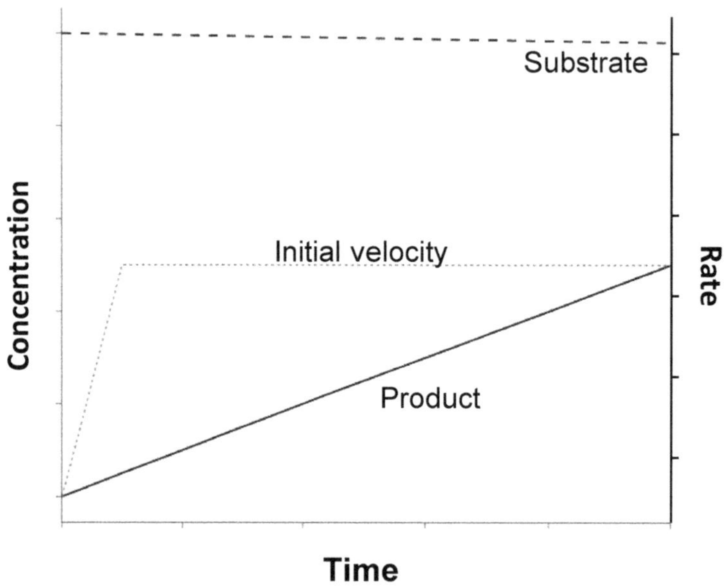

FIGURE 9.6. Unlike typical reaction progress curves that exhibit continually changing rates (e.g., Figure 9.1), the steady-state condition yields a constant reaction velocity. Under this condition substrate consumption is minimal, and therefore the concentration of ES complex is constant, giving rise to the constant measured velocity. This state persists until sufficient substrate is consumed or product is generated to influence the forward or reverse reaction rates and the ES concentration.

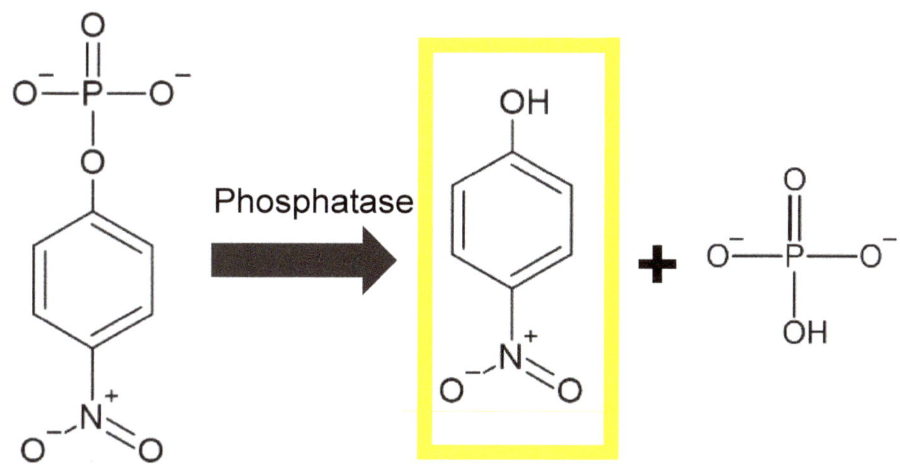

FIGURE 9.7. *Para*-nitrophenyl phosphate (left) is a common substrate for measuring phosphatase enzyme activity. Phosphatases hydrolyze the ester bond between the phosphate and the phenyl ring, producing the products *para*-nitrophenol and inorganic phosphate. *Para*-nitrophenol has a deep yellow color and absorbs light strongly at 405 nm, providing a convenient spectroscopic method for monitoring reaction progress and measuring product concentration.

A SIMPLE ASSAY FOR PHOSPHATASE ACTIVITY

The chemical *para*-nitrophenyl phosphate (pNPP) is commonly used to assay the activity of a broad range of phosphatase enzymes. Hydrolysis of its phosphoester bond (P-O-C) by phosphatases converts pNPP into *para*-nitrophenol (pNP) and inorganic phosphate (Figure 9.7). pNP is a yellow compound that absorbs light strongly at 405 nm under basic conditions (which cause deprotonation of the hydroxyl group), and therefore the reaction can be followed spectrophotometrically by measuring the increase in absorbance at 405 nm wavelength. pNPP is inexpensive, stable, and safe. Although pNPP is not a natural physiological substrate for phosphatases, it is commonly used to study phosphatase enzymes in the lab because of its convenient assay properties.

OBJECTIVES

The goal of today's lab is to determine if your purified protein has the enzyme activity you hypothesized back in Week 1. Remember that your hypothesis was based on the presence of a domain homologous to the protein tyrosine phosphatase family, specifically the Cdc14 phosphatases within this family, and your literature search on known functions of Cdc14. Today you will set up and perform a generic assay for measuring phosphatase enzyme activity (with pNPP as the substrate) and will compare the wild-type enzyme to the predicted catalytic mutant generated by site-directed mutagenesis.

If you detect the predicted enzyme activity, you will then establish assay conditions to be used in subsequent weeks for steady-state kinetic assays to explore such things as catalytic rates, substrate specificity, and inhibition of your enzyme by small molecules.

LAB INSTRUCTIONS

PART 1. PREPARE ENZYME STOCKS FOR ASSAYS

1. Retrieve your final concentration measurement for your FoCdc14 enzyme preparation in units of μM (you calculated this from your Bradford assay data in the Week 6 postlab question set). Make sure everyone in your group obtained the same value, and confirm with the instructor or the teaching assistant that your calculation is correct.
2. Obtain the same measurement for your partner group's preparation. You have two enzyme preparations to compare today: wild-type and predicted catalytic mutant. One preparation will be your own (either wild-type or mutant), and the other will be borrowed from another group.
3. Record the protein concentration values in your notebooks where they can be easily found. You will use these values again in upcoming weeks.
4. Make 100 μl of a 10x working dilution of each FoCdc14 preparation in reaction buffer for Part 2 (remember, 10x means 10 times the desired final concentration). This should be a 10 μM solution but may need to be altered based on the results of the experiments below. Additional protein dilutions will need to be made for Part 4.

PART 2. TESTING FOR PHOSPHATASE ACTIVITY AND CONSTANT REACTION VELOCITY

Now you will test your original hypothesis that your protein has phosphatase activity. You will measure activity in real time over the course of 20 minutes with an initial set of reaction conditions and determine if the graph of product formation versus time is linear. The comparison of wild-type FoCdc14 and its predicted catalytic mutant will reveal not only if FoCdc14 has the hypothesized enzyme activity but also if it functions by the same mechanism as other characterized Cdc14 enzymes. Assessing the relationship between reaction velocity and time is important for establishing steady-state conditions, which will be needed in the next couple of weeks. In the steady state there is a constant reaction velocity, and we want that to be maintained for at least 20 minutes. Make sure you understand all the reasons why reaction velocity could change over time.

1. You will run the three reactions indicated in Table 9.1 in triplicate (nine total wells in a single row of the microplate in the order blank 1, wild-type 1, mutant 1, blank 2, wild-type 2, mutant 2. . .). First, add 105 μl reaction buffer to each well, followed by 30 μl of substrate (pNPP). Mix each well by pipetting up and down a few times after adding the pNPP, but be careful to not introduce bubbles into the mixture (only depress the piston to the first stop position).
2. Next, transfer 100 ul reaction buffer and 100 μl of the 10x stocks of wild-type and mutant enzyme prepared in Part 1 to the three empty wells in the row just beneath your replicate 1 reaction wells. This is more volume than you need but will facilitate transferring the correct amount to the reaction wells.

TABLE 9.1. *Reaction Setup for Measuring Phosphatase Activity*

REACTION	REACTION BUFFER (μL)	100 mM pNPP (μL)	10X ENZYME STOCK (μL)
Blank	120	30	0
Wild-type enzyme	105	30	15 (for 1 μM final)
Mutant enzyme	105	30	15 (for 1 μM final)

3. Start the plate reader's kinetic measurement method for absorbance at 405 nm (the teaching assistant or the instructor will help with this). It will stop automatically, with the door open and the plate holder extended with a prompt to place the plate in the tray.
4. Set a multichannel pipettor to 15 μl and attach three tips to adjacent dispensing heads on it. Draw up 15 μl of the buffer and enzymes and add to the three replicate 1 wells to initiate the reactions. Repeat this two more times without changing tips to rapidly start the replicate 2 and 3 reactions. For each replicate, mix by pipetting up and down to the first stop position on the pipettor. The mixing should be done quickly because the reaction starts as soon as your enzyme is added.
5. Multiple groups may be asked to load their reactions on the same plate to save time. Reactions should only be started when all groups using the plate are ready. Place the microplate in the plate reader after all reactions have been started, select the wells to read, and click OK to initiate the kinetic read.
6. The data collection will stop automatically after 20 minutes. The final absorbance values will be exported for you in Excel, but you can also look at the graphs in real time on the plate reader software to ensure that you have used enough enzyme to get a robust absorbance signal and to conclude if the reaction velocity for the wild-type enzyme remains constant over the 20 minutes (linear increase in absorbance).
7. While data is collecting, you should begin preparing for Parts 3 and 4 below. Note that if you are not getting sufficient absorbance signal with wild-type enzyme you may have to repeat this assay with higher enzyme concentration (the instructor will let you know if this is the case).

PART 3. GENERATE A *PARA*-NITROPHENOL (pNP) STANDARD CURVE AND USE IT TO MEASURE SUBSTRATE CONSUMPTION

Just like for the Bradford assay, you will need a good standard curve relating reaction product concentration (pNP) and absorbance. This curve will be saved and used in future weeks as well.

1. Make a set of standard samples of the product, pNP, at the following concentrations, starting with the provided 5 mM stock solution: 0, 12.5, 25, 50, 100, 150, 200, and 300 μM.
   A. Read through each step carefully and then make all your calculations before beginning.
   B. An easy way to create a set of standards at different concentrations is by using serial dilution steps. For example, you can use the 200 μM stock as the source to create your 100 μM stock simply by taking 500 μl of the 200 μM solution and mixing with 500 μl buffer. Then that new 100 μm stock is the source for making the 50 μM stock by the same procedure. Repeat this until you've made all of the twofold dilutions.
   C. Note that you will need to make both 300 μM and 200 μM pNP dilutions directly from the 5 mM stock to generate all of the standards by twofold serial dilution (300 → 150 and 200 → 100 → 50 → 25 → 12.5).
2. Pipet 150 μl of each pNP solution into microplate wells in triplicate, one set per row. You should have three separate rows of pNP samples ranging from 0 to 300 μM. The instructor or the teaching assistant will help you collect absorbance readings at 405 nm on these with the plate reader.
3. Export the data to Excel and use your flash drive to transfer the file to your lab computer. Open the file in Excel and use the data to create a graph of average absorbance as a function of pNP concentration, showing standard deviation error bars.
   A. You will need to first subtract the background absorbance measured with 0 μM pNP from all the other absorbance values.
   B. Include 0,0 as a data point in the standard curve.

4. Apply a linear regression to the graph (force the line through 0) and include the equation. This is your standard curve. The linear fit equation will be used to convert absorbance to product concentration in all of your pNPP enzyme assays in Part 4 and subsequent weeks.
5. Before initiating Part 4, use your standard curve to calculate the concentration of product produced in your wild-type enzyme assay from Part 2. Subtract the average blank reaction absorbance from the average final absorbance reading for the wild-type enzyme at 20 minutes. Plug this value into the standard curve equation to convert to product concentration (equal to substrate consumed).
   A. Is less than 10% of the starting substrate consumed?
   B. If so and if your reaction plot from Part 2 looks linear over the entire 20 minutes, then you have acceptable steady-state conditions for the future kinetic analyses and inhibition studies.
   C. If not or if your reaction plot from Part 2 does not maintain linearity for 20 minutes, review your results with the instructor to make adjustments to reaction conditions prior to Part 4. You will likely need to reduce your enzyme concentration.

PART 4. TEST IF REACTION RATE IS PROPORTIONAL TO ENZYME CONCENTRATION

For simple enzymes under steady-state conditions, the measured reaction velocity should be proportional to the enzyme concentration—that is, the rate per unit of enzyme should be constant. It is important to establish this and know that changes in enzyme concentration that might be needed in upcoming weeks will not affect kinetic measurements. To test the relationship between reaction rate and enzyme concentration, set up the four reactions illustrated in Table 9.2 in triplicate (12 reactions total). The table is set up for final FoCdc14 concentrations of 4, 2, 1, and 0.5 μM. If the enzyme concentration had to be adjusted in Part 2 to get good activity, then adjust here as well.

1. For each replicate, first mix enough reaction buffer and substrate together for five reactions. This is called a "Master Mix." Then, add 135 μl of this master mix to four consecutive microplate reaction wells in a single row of the microplate. Make sure there are no large bubbles. Repeat for the other two replicates, adding 135 μl of master mix to the corresponding wells in the two rows below the replicate 1 row.
2. Transfer 100 μl of each 10x Cdc14 solution to the appropriate well in the row below the replicate reaction rows.
3. Get the plate reader set up and start the kinetic measurement at 405 nm for 20 minutes. The instructor or the teaching assistant will help with this. Pause at the point where you are asked to insert the plate into the tray.
4. Using a multichannel pipettor, simultaneously transfer 15 μl from each of the 10x enzyme wells into the corresponding reaction wells, mix well by pipetting up and down (but carefully so as not to generate bubbles—again, only go to the first stop position). Without changing tips, repeat for replicates 2 and 3.

TABLE 9.2. ***Reaction Setup for Measuring Dependence on Enzyme Concentration***

REACTION	REACTION BUFFER (μL)	100 mM pNPP (μL)	10X ENZYME STOCK
1	105	30	15 μl 40 μM
2	105	30	15 μl 20 μM
3	105	30	15 μl 10 μM
4	105	30	15 μl 5 μM

5. Immediately put the plate in the plate reader tray and click OK to start the measurements.
6. Export the data to an Excel file and transfer to your computers using your flash drive.

DATA ANALYSIS AND SUBMISSION

A single Excel File should be submitted to the course learning management system containing separate pages for each of the following parts and providing all of the items listed. As before, use Excel functions to show work and state conclusions when requested on the relevant pages.

FOR PART 2:

- Open your exported Excel data file.
- Using Excel functions, calculate the average absorbance for the three trials of the blank, wild-type, and mutant enzyme assays.
- Create a scatter graph with the average absorbance values plotted as a function of time. Note: The exported plate reader data may have time in units of h:mm:ss (hours:minutes:seconds). To convert these values to normal numbers in units of minutes, multiply each cell by 1,440.
- Perform linear regression on each. You should expect to see measurable activity with the wild-type enzyme (steep positive slope) with a constant, linear increase in absorbance over the 20 minutes.
- The linearity of the graph is important to consider because it indicates whether the steady-state condition exists throughout. If the data points lose linearity, first check if the later absorbance values exceed the linear standard curve range. If they do there is too much activity, and the enzyme concentration must be lowered.
- The mutant reaction plot should be almost flat, similar to the blank (representing no activity) if the hypothesis that your protein is a true Cdc14 ortholog is correct. Evaluate this plot compared to the wild-type enzyme plot and come to a conclusion about the effect of the mutation and whether these results support your original hypothesis.
- Include your calculation of substrate consumption for the wild-type enzyme and conclusion described in Part 3, item 5, above.

FOR PART 3:

- Include the raw absorbances from the plate reader, your calculated averages and standard deviations, and your final standard curve graph with error bars (standard deviations) and linear regression, including the standard curve equation on this sheet.

FOR PART 4:

- Open your exported Excel data file.
- Create a scatter graph of all individual trials (absorbance vs. time) and assess the linearity of each. As described above, convert time into standard number format (minutes) first if needed.
- Use the Excel slope function to determine the slope (absorbance/minute) of each reaction, using only the linear portions of the data.

- Convert all the slopes from each reaction to [product]/min using the standard curve equation (plug slopes in as y values).
- Calculate average rates for each enzyme concentration with standard deviations.
- Generate a graph showing observed reaction rate as a function of enzyme concentration, with error bars.
- Perform linear regression on this graph. For simple enzymes, reaction rate should be proportional to enzyme concentration. Conclude whether or not the reaction mechanism for your enzyme is independent of enzyme concentration.
- Calculate total substrate consumption after 20 minutes (in percent) for each enzyme concentration and note if any surpass 10%.
- Include a final conclusion on whether the steady state is maintained for at least 20 minutes under your reaction conditions.

10. STEADY-STATE ENZYME KINETICS

BACKGROUND AND THEORY

THE MICHAELIS-MENTEN EQUATION

A simple one-substrate enzyme reaction is described by the scheme in Figure 10.1, where E is enzyme, S is its substrate, ES is the enzyme-substrate complex, and P is product. Each step in the reaction is reversible, in principle, so there are four distinct reaction steps.

Early in the reaction where P is negligible, the reverse reaction (step −2) can be ignored. The other steps in the reaction can be written as rate equations (Figure 10.2) with the equivalency $v_1 = v_2 + v_{-1}$ defining the steady-state condition (it gives rise to constant ES concentration).

The terms in the equation $v_1 = v_2 + v_{-1}$ can be rearranged to give a function describing the initial velocity of an enzyme-catalyzed reaction at steady state as a function of substrate concentration (see lecture for additional detail). This equation was first described by the scientists Leonor Michaelis and Maud Menten[1] and is often called the Michaelis-Menten equation (Figure 10.3). This equation is a hyperbolic function. The relationship between reaction velocity and substrate concentration is hyperbolic because of the phenomenon of enzyme saturation, which occurs at high substrate concentrations relative to enzyme (see lecture for additional information on this).

FIGURE 10.1. The simplest possible enzyme-catalyzed reaction scheme has two reversible steps: the binding of enzyme (E) and substrate (S) to form the enzyme-substate complex (ES) and the conversion of S to product (P) and release to regenerate free enzyme. ES can also dissociate back to free E and S (step −1), and E and P can associate to form an EP complex to initiate the reverse reaction (step −2).

$$v_1 = k_1[E][S] \qquad v_2 = k_2[ES]$$

$$E + S \rightleftharpoons ES \longrightarrow E + P$$

$$v_{-1} = k_{-1}[ES]$$

At steady state:

$$v_1 = v_2 + v_{-1}$$

OR

$$k_1[E][S] = k_2[ES] + k_{-1}[ES]$$

FIGURE 10.2. When set up in the lab, the enzyme scheme from Figure 10.1 can be simplified further to ignore the reverse reaction because there is no product initially. The three remaining steps (1, –1, and 2) have associated rate equations as indicated. By definition, in the steady state v_1 will be equal to $v_2 + v_{-1}$, as this results in constant [ES]. This equivalency can also be written using the full rate equation terms from each step.

The Michaelis-Menten Equation

$$v_0 = \frac{k_2[E_T][S]}{(K_m + [S])} \qquad k_2[E_T] = V_{max} \qquad v_0 = \frac{V_{max}[S]}{(K_m + [S])}$$

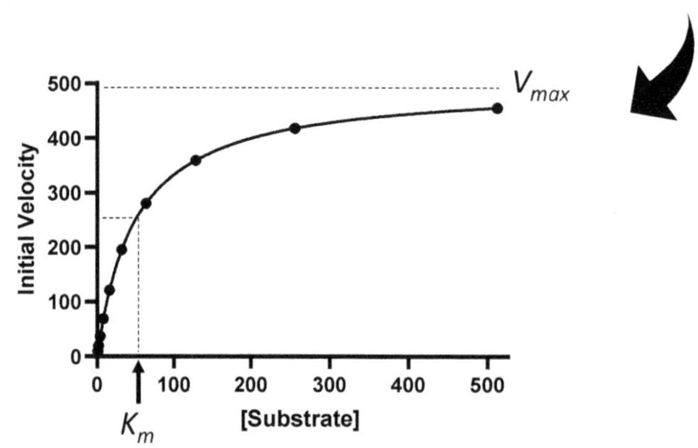

FIGURE 10.3. Rearrangement of the equivalency shown in Figure 10.2 to represent the observed (initial) velocity (v_0) under steady-state conditions gives rise to the famous Michaelis-Menten equation, which is a hyperbolic function relating v_0 to substrate concentration. The asymptote of the hyperbola can be expressed either as V_{max}, which is the maximum initial velocity of the experimental reaction at saturating substrate concentration or as k_2[ET]. k_2 is also called k_{cat}, the turnover rate of the enzyme. The relationship between V_{max} and k_2 (k_{cat}) is shown. The other variable parameter is K_m, defined as the substrate concentration that results in an initial velocity equal to half of V_{max}.

THE STEADY STATE KINETIC PARAMETERS k_{cat} AND K_m

Hyperbolic functions have the general form y = ax/(b+x), with a and b as the variable parameters that define its shape. In the Michaelis-Menten equation the "a" parameter has a special name, V_{max}, which is the maximal reaction velocity (asymptote) that the hyperbolic curve approaches. V_{max} represents the maximum rate of catalysis when all enzyme active sites are continuously occupied with substrate molecules, a situation called enzyme saturation. V_{max} will vary proportionally with enzyme concentration. Increasing the enzyme concentration twofold will give a twofold increase in measured V_{max}. Thus, reporting V_{max} also necessitates reporting the enzyme concentration used. A more useful parameter is V_{max} divided by the enzyme concentration, called the k_{cat}, or catalytic turnover rate. k_{cat} is the number of reactions catalyzed by a single enzyme molecule per unit of time (usually per second). Note that k_{cat} is equivalent to k_2 in Figures 10.2 and 10.3.

The "b" parameter in the Michaelis-Menten equation also has a special name, K_m, which is operationally defined as the substrate concentration that yields a reaction velocity equal to half of V_{max}. For many enzymes (but not all), K_m is a reasonable approximation of the affinity for the substrate. Thus, you can think of k_{cat} as a measure of how fast an enzyme catalyzes a reaction and K_m as a measure of how tightly the enzyme binds its substrate. These two properties are very useful to know when characterizing enzymes, and you will see examples of why in the upcoming weeks. The goal of steady-state kinetics experiments is often to measure these two parameters.

MEASURING k_{cat} AND K_m IN THE LAB

How does one determine k_{cat} and K_m from enzyme assays? The experiment must be designed in a specific way to generate the hyperbolic initial velocity versus substrate concentration graph. Then, curve-fitting approaches can be used to extract the k_{cat} and K_m values. The key element of the experimental design is measuring initial velocities at varying substrate concentrations that start well below the K_m and end well above it. A good target range is tenfold below to tenfold above the K_m. For each of the reactions at a different substrate concentration it is essential to maintain the steady-state condition to ensure that the initial velocity is measured ([E] << [S] and <<10% substrate consumed). This reaction design will produce a suitable hyperbolic curve for accurate estimation of k_{cat} and K_m by curve fitting.

Historically, before the advent of computers the most common method for estimating k_{cat} and K_m from data generated by velocity versus substrate concentration experiments was to convert the Michaelis-Menten function to a linear form so that linear regression could be applied to estimate k_{cat} and K_m. There are several ways to do this, with the method of Lineweaver and Burk (plotting the reciprocal of the y and x values) being particularly popular. However, linearization has drawbacks, and with computers available there is no longer a need for linearization of the Michaelis-Menten function. Instead, the hyperbolic data can be directly analyzed by nonlinear least squares curve fitting, which you will apply to your own data this week. In the least squares method, the variable equation parameters are systematically varied by a computer. For each combination, the computer generates a theoretical curve and data points. For each x value, the computer calculates the difference between the measured and theoretical y values, squares them (to make everything positive), and sums them (Figure 10.4). The computer then finds the parameter combination that gives the lowest value for the summed squared differences. These are called the best fit parameters, and the line defined by these parameters is the best fit curve.

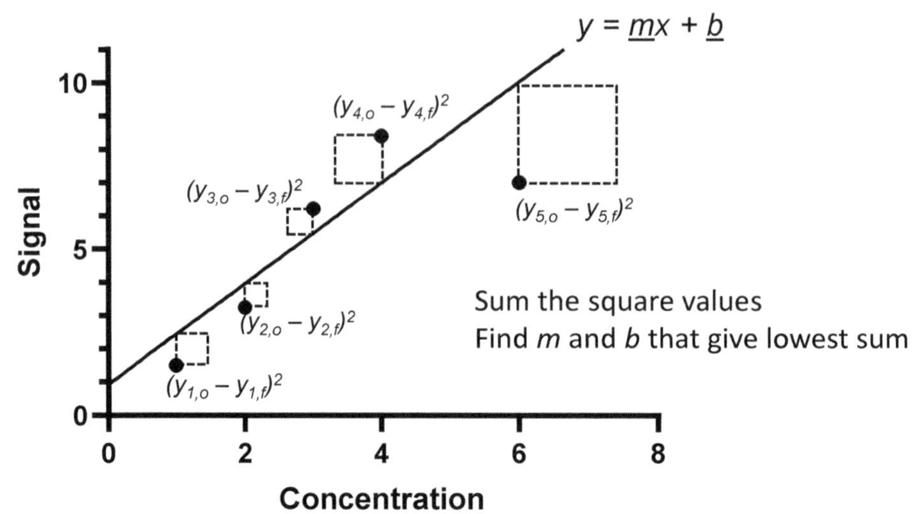

FIGURE 10.4. In least squares curve fitting the variable parameters of a fitting function are systematically varied to find the minimum sum of squared differences between the observed (o) and fit (f) y values. In this example, the linear function y=mx+b is used, and a single fit line is shown. For each x there is a measured y value and a y value from the fit line. The differences between these are squared (to make all values positive) and summed. The parameters m (slope) and b (y-intercept) that define the fit line are varied until the ones that produce the smallest sum of squared differences is found. This is all done by computer.

OBJECTIVES

Kinetic analysis can be used to determine many characteristics of an enzyme such as its affinity for substrates, its maximal rate of catalysis, and the effectiveness of an enzyme inhibitor. Last week you learned about the steady-state condition that is commonly used to assess these properties of an enzyme. The goal of today's lab is to measure the steady-state kinetic parameters k_{cat} and K_m for your enzymes and compare them to results published for the related enzyme from *Saccharomyces cerevisiae* using the same common phosphatase substrate, pNPP, that you used last week. This will tell you if your FoCdc14 protein possesses enzymatic properties similar to a well-characterized member of this enzyme family. This assay will also form the basis for exploring inhibition of your enzymes in Week 12.

A secondary objective is to obtain a working understanding of nonlinear least squares curve fitting, which you will use to derive best fit values of k_{cat} and K_m from your kinetics data.

LAB INSTRUCTIONS

PART 1. GENERATE A VELOCITY VERSUS SUBSTRATE CONCENTRATION PLOT TO DETERMINE STEADY-STATE KINETIC PARAMETERS

To measure the steady-state kinetic parameters K_m and k_{cat} for your enzyme, you will determine the rate of dephosphorylation of pNPP as a function of pNPP concentration. You will then fit your data with the Michaelis-Menten equation in Excel.

You will use the exact same reaction setup for absorbance measurements in the plate reader that you used last week to evaluate enzyme concentration dependence of your Cdc14 activity. This time, though, reactions will be started by adding substrate using the multichannel pipettors and absorbance at 405 nm will be measured over time. You have your previous standard curve of the product pNP that you will need to convert absorbances into product concentrations.

1. Set up your reactions and measure the product formation according to the guidelines in Table 10.1. *Please read everything carefully and plan thoroughly before beginning.*
2. You can set up your reactions directly in the microplate wells, but you must avoid getting lots of bubbles in the wells. You will need three columns of wells, one for each of the three replicates, and a fourth column to load your 10x pNPP stocks. Set up your plates as follows:
   Columns 1, 2, 3: Reactions for trials 1, 2, 3, respectively
   Column 4: 250 μl of all of the 10x pNPP stocks in order from least to greatest concentration
   You may be asked to share a plate with another group, in which case your reactions may be loaded in different number columns, but the organization should be the same.
3. Make your series of 10x pNPP stocks. You will need eight total dilutions, one for each of the final concentrations listed in Table 10.1. Dilute the pNPP in 25 mM DEA pH 9.8. The 1000 mM stock solution is your 10x stock for the 100 mM pNPP reaction. You just need to make twofold dilutions from that to get the remaining 10x substrate stocks. Start by combining 500 μl 1000 mM pNPP with 500 μl DEA solution and vortexing to mix. Then repeat to get each subsequent dilution. Finally, you will need to make 500 μl of the 10x stock for the 75 mM reaction (750 mM) separately from the 1000 mM stock (do this calculation yourselves).
4. Make your 5x enzyme stock in reaction buffer next (usually 10 μM is appropriate, but this may need to be adjusted based on results in Part 4 from last week), with enough volume for four full trials so that you have some extra left over.
5. Then, make a master mix of buffer + enzyme for each of the three trials. A master mix is a single mix that includes all of the shared ingredients for all reactions. It is then aliquoted to individual reaction wells. Master mixes save time and reduce variability. Always calculate your master mix volumes for one more reaction than you need so that you don't run out while aliquoting (i.e., for the eight reactions in each trial, make enough master mix for nine reactions).
6. Aliquot 135 μl enzyme/buffer master mix to the reaction wells in columns 1, 2, and 3.
7. Transfer 250 μl each pNPP stock to a well in column 4 (in order of increasing concentration from top to bottom).

TABLE 10.1. *Reaction Setup for Michaelis-Menten Kinetics Experiment*

REACTION—FINAL [pNPP]	REACTION BUFFER (μL)	10X SUBSTRATE (μL)	5X ENZYME STOCK (μL)
1.55 mM	105	15	30
3.12 mM	105	15	30
6.25 mM	105	15	30
12.5 mM	105	15	30
25 mM	105	15	30
50 mM	105	15	30
75 mM	105	15	30
100 mM	105	15	30

8. To initiate reactions, take your plate to the microplate reader where the teaching assistant or the instructor will start a new read at 405 nm. The software will prompt you to load your plate. At this point, use the multichannel pipet to transfer 15 μl each pNPP stock sequentially into columns 1–3. You do not need to change tips in between columns. For each column, mix by pipetting up and down several times to the first pipettor stop position (to avoid bubbles). If another group is on the same plate, they should start their reactions immediately after.
9. Immediately transfer the plate into the reader and hit OK to close the plate drawer and initiate the read. It is programmed to read for 15 minutes, recording the absorbance after each minute. Export the data to Excel, save to your group's flash drive, and transfer to your lab computers for analysis and graphing.

PART 2. USE NONLINEAR LEAST SQUARES CURVE FITTING TO DETERMINE V_{max} AND K_m

1. Open the exported data file in Excel on your computer. You have been provided with an Excel template file to facilitate calculations and curve fitting. Open that Excel template on your computer as well. The instructor or the teaching assistant will walk you through the template.
2. Copy the raw absorbance values from your exported data file into the first sheet titled "Raw Absorbances" in the analysis template.
   A. Create scatter plots of absorbance versus time for all trials at each pNPP concentration to determine if initial velocity is maintained throughout the reaction. A plot should autogenerate for the first trial at 1.56 mM pNPP. You just need to change the data used in the graph to see plots for each of the other trials.
   B. If any of the plots level off over time, you will need to exclude the data points that are nonlinear.
   C. Use the Excel Slope function in row 21 to calculate absorbance/min for each reaction using only the linear data points.
3. Copy your standard curve data from last week into the second sheet of the analysis template and re-create your standard curve.
4. To calculate initial velocities for each reaction, go to the third sheet in the analysis template titled "Conversion to pNP rates."
   A. Copy the slope value from your standard curve equation into cell B1.
   B. Copy all the absorbance/min values from row 21 of sheet 1 into row 5.
   C. Use an Excel function and your standard curve slope (cell B1) to convert the values in row 5 into product formation rates (μM pNP/min) in row 6.
5. Transfer all of your μM/min rates for your three trials into the appropriate cells of the fourth sheet titled "Curve Fitting."
6. The template will automatically convert your rates into μM/sec units and graph the reaction velocities as a function of substrate concentration for each individual trial.
7. The instructor will then walk you through how to use the Excel Solver add-in to perform least squares curve fitting and determine the best fit V_{max} and K_m values for each of your three trials and the averaged data.
8. Record the best fit V_{max} and K_m values for each trial and the averaged data in the yellow highlighted section of rows 25 and 26. Convert your V_{max} values to k_{cat} values in row 27 by dividing them by your enzyme concentration (in μM). This gives (μM pNP/sec)/μM enzyme. The μM units cancel out, leaving the standard k_{cat} units of sec^{-1} (1/sec). Average V_{max}, k_{cat}, and K_m values with standard deviations will be automatically calculated in the orange highlighted sections of rows 25–27.
9. For your final reports, you should show the graph of the averaged data with its line of best fit from this analysis template ("Average V vs. S" graph). And you should report the average k_{cat} and K_m values with standard deviations from the orange highlighted section as well.

DATA ANALYSIS AND SUBMISSION

Submit your completed Excel analysis template file to the course learning management system with all data analysis, graphing, curve fitting, and calculations completed as described above. Make sure this file is saved to your group data storage folders as well.

Compare your steady-state kinetic parameters with those previously reported from *Saccharomyces cerevisiae* Cdc14 on the same pNPP substrate (there are a couple of papers that have these values in them).[2] Are they mostly consistent activities, or are there significant differences? On the curve-fitting sheet of your Excel analysis file, provide a description of how your enzyme's steady-state kinetic parameters compare to those determined previously for *Saccharomyces cerevisiae* Cdc14 with pNPP.

Make sure you have all data, data analysis, and experimental steps thoroughly recorded in your notebooks. You will need to refer back to these assays when you study substrate specificity and enzyme inhibition later.

11. SUBSTRATE SPECIFICITY

BACKGROUND AND THEORY

PRINCIPLES OF MOLECULAR RECOGNITION

In biological systems, proteins must be able to identify the appropriate ligands, substrates, cofactors, interaction partners, etc. out of an enormously complex mixture of different molecules. Once found, they must bind these molecules tightly enough to elicit the appropriate biochemical response or output. The ability of proteins to recognize and bind particular molecular structures is called specificity. What is the basis for the incredible binding specificity that proteins display toward substrates and interacting molecules? Among enzymes with multiple substrates, what distinguishes an excellent substrate from a relatively poor substrate and from a molecule that is not processed as a substrate at all?

Specific binding between biomolecules usually involves the aggregate effect of multiple weak (noncovalent) chemical bonds. These bonds require a specific 3D arrangement of complementary chemical groups on the surfaces of the interacting molecules (Figure 11.1). Thus, specific molecular interactions are dependent on (1) complementary 3D shapes of the interacting molecules, (2) the *presence* of complementary chemical groups that can form noncovalent bonds, and (3) the *absence* of noncomplementary or clashing chemical groups that would disfavor binding. Binding affinity is determined largely by the number and strength of these weak chemical bonds that exist between the interacting molecules.

As described in more detail in the lecture, the types of noncovalent chemical bonds that contribute to specific molecular recognition in biological systems (Figure 11.2) are as follows:

- Hydrogen bonds. These occur between hydrogens covalently bound to electronegative atoms (oxygen and nitrogen in proteins, for example) and other electronegative atoms. The hydrogen obtains a partial positive charge, or dipole, because of the electronegative atom pulling electron density away from it. This allows for a weak ionic interaction with the partial negative charge (dipole) on other electronegative atoms.
- van der Waals interactions. These are weak interactions between molecular dipoles of opposite charge. It is a similar principle to hydrogen bonds, but is not restricted specifically to hydrogen atoms. All atoms can attain temporary or permanent induced dipoles that can engage in van der Waals interactions. These interactions are very weak and occur only at very short distances between adjacent atoms, but collectively they can contribute significantly to such things as ligand and substrate binding and to protein folding.
- Hydrophobic interactions. Nonpolar chemical groups have a natural affinity for each other based on their shared tendency to repel, or exclude, water molecules. Thus, in an aqueous environment, nonpolar groups will preferentially associate with each other rather than the polar solvent molecules. This is thermodynamically favored

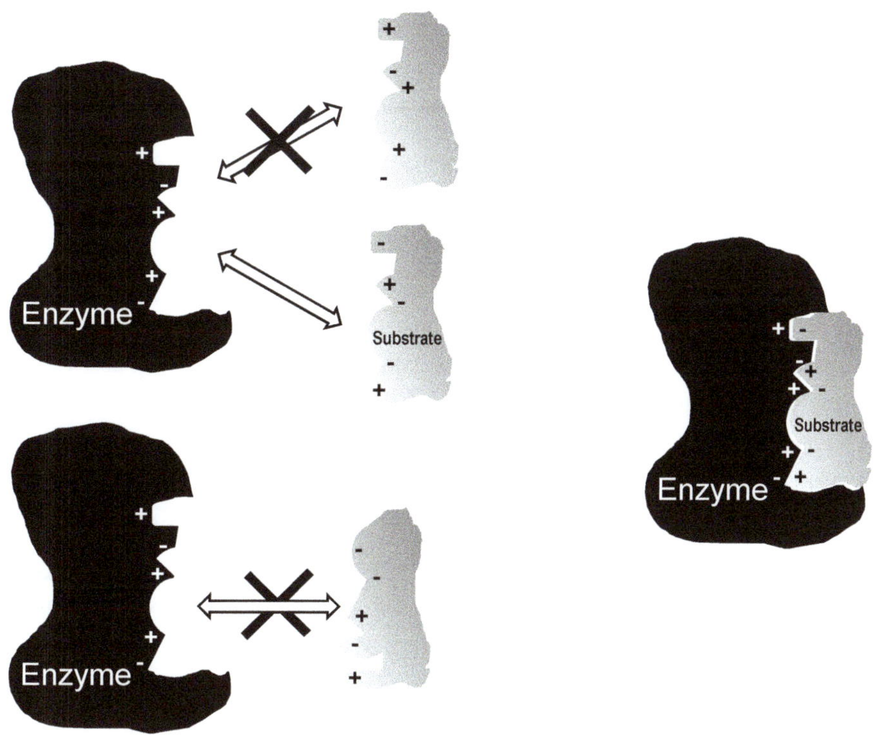

FIGURE 11.1. Specific protein-ligand interactions in biology, such as those of enzymes with their substrates, are driven by shape complementarity and the juxtaposition of functional groups capable of engaging in weak noncovalent bonds. At top left, shape complementarity is not sufficient for binding, as noncompatible functional groups (+ and – charges) will prevent interaction. At bottom left, compatible functional groups cannot drive interaction alone if shape complementarity does not exist. The productive enzyme-substrate interaction at right requires both.

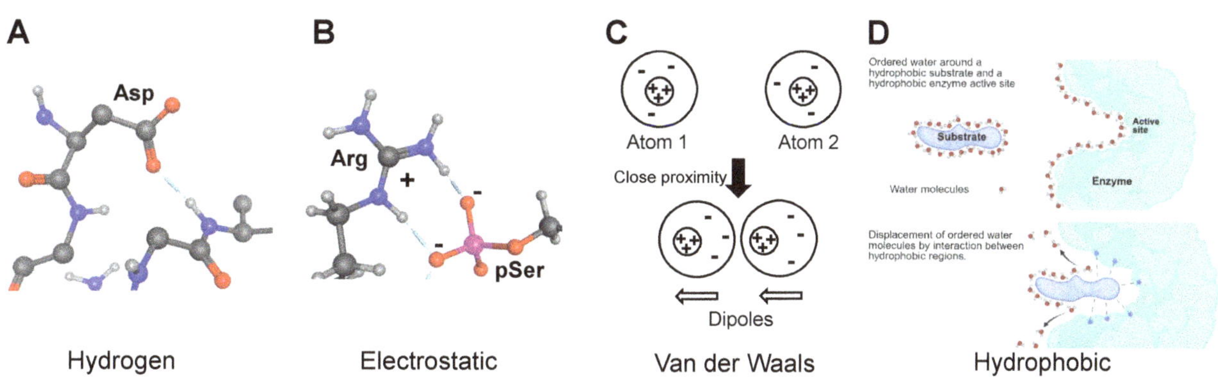

FIGURE 11.2. Specific binding of ligands by proteins are driven by hydrogen bonding (A), electrostatic bonds (B), van der Waals interactions (C), and hydrophic interactions (D). In A and B, red spheres are oxygen, blue spheres are nitrogen, large gray spheres are carbon, and small gray spheres are hydrogen. The magenta atom in B is phosphorus, and pSer indicates a phosphorylated serine amino acid. RCSB PDB ID: 5XW5 was used to generate images in panels A and B. Panels A and B were generated with Molecular Operating Environment. Panel D was created in BioRender, Hall, M. (2026) https://BioRender.com/t90z6a1.

because it increases entropy, but it is not an actual bond. It is what causes oil and water to separate when mixed and can also encourage protein-ligand interactions by the same principle.

- Ionic (or electrostatic) bonds. Ionic bonds are based on the attractive force between positive and negative charges. At physiological pH, chemical groups that carry a positive charge (e.g., the side chains of lysine and arginine) can form ionic bonds with those carrying a negative charge (e.g., the side chains of aspartic and glutamic acid).

CATALYTIC EFFICIENCY AS A MEASURE OF SUBSTRATE BINDING

Based on the principles described above, enzymes typically recognize a small number of specific substrate molecules. Studying substrate recognition by enzymes in the lab is challenging using assays that measure binding directly, because once enzymes bind a substrate, they rapidly convert it to product. However, under the appropriate conditions, enzyme assays can be used to compare the binding of different substrate molecules. Last week you learned that two features of an enzyme can be measured in steady-state kinetic experiments, K_m and k_{cat}, where K_m often approximates the affinity for the substrate and k_{cat} represents the rate at which a bound substrate molecule is converted to product. This week you will learn that the ratio k_{cat}/K_m, often called the catalytic efficiency or specificity constant, is a useful measure of substrate binding and is often used to compare the binding of different substrate molecules to a given enzyme. This can help identify optimal substrates or define general principles of an enzyme's substrate specificity.

As you learned last week, determining k_{cat} and K_m requires measuring reaction rate as a function of increasing substrate concentration. This takes significant time and effort and can be cumbersome when evaluating large numbers of candidate substrate molecules. There is an easier way to estimate k_{cat}/K_m without directly calculating the individual k_{cat} and K_m parameters from velocity versus substrate concentration curves as you did last week. In a standard velocity versus substrate concentration curve, the initial slope (at low substrate concentrations where there are no saturation effects) is proportional to k_{cat}/K_m (Figure 11.3), and individual measurements at substrate concentrations well below the K_m are reasonable estimates of the catalytic efficiency for many enzymes.

Such measurements are useful when, for example, screening large libraries of candidate substrates where calculating individual k_{cat} and K_m values with a full velocity versus substrate concentration curve is impractical or when conditions of enzyme saturation cannot be achieved in steady-state kinetic assays.

AN ASSAY FOR MEASURING PHOSPHATE RELEASE

At low pH, inorganic phosphate and molybdate can form a complex with a synthetic dye called malachite green that alters its spectral properties, producing a deep green color and a strong absorbance peak in the 600–660 nm wavelength range (Figure 11.4). This property of malachite green can be used to assay for just about any enzyme that generates free phosphate as a product of its catalytic reaction (phosphatases, ATPases, etc.). You will use it to assay FoCdc14 activity on biologically relevant phosphopeptide substrates to characterize its natural substrate specificity.

OBJECTIVES

The primary objective for this week is to characterize the substrate specificity of FoCdc14 by measuring k_{cat}/K_m for a small set of phosphopeptide substrates and comparing it to the previously characterized specificity

$$v_0 = \frac{k_{cat}[E_T][S]}{(K_m + [S])} \xrightarrow{\text{At very low [S]}} v_0 = \frac{k_{cat}[E_T][S]}{K_m}$$

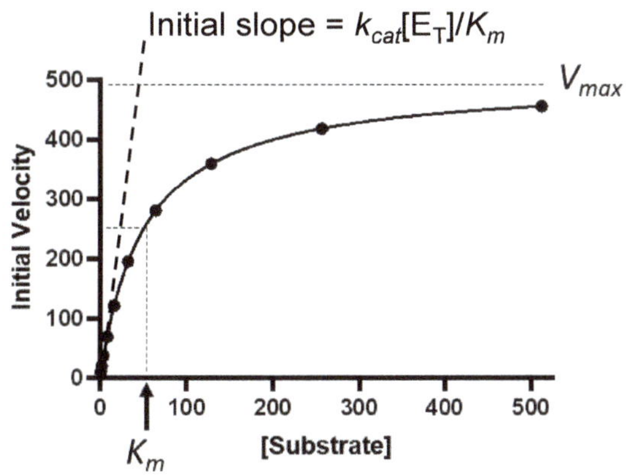

FIGURE 11.3. At very low substrate concentrations the Michaelis-Menten function is linear, with a slope = $k_{cat}[E_T]/K_m$. This is because the low substrate concentration relative to K_m makes a negligible contribution to the denominator. As a result, the catalytic efficiency value, k_{cat}/K_m, can be conveniently determined by conducting assays at low substrate concentration without the need to establish full velocity versus substrate concentration plots and individually measure k_{cat} and K_m.

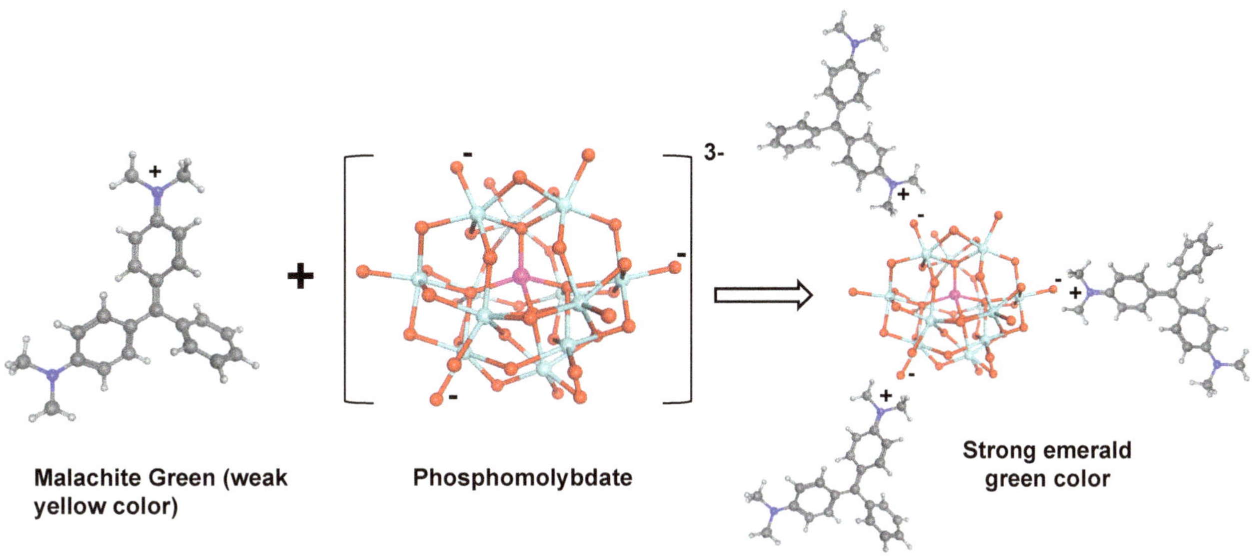

FIGURE 11.4. When complexed with molybdate and inorganic phosphate through electrostatic bonds, the dye malachite green becomes an emerald green color, with strong light absorbance at 640 nm. This phenomenon is the basis for assays to detect inorganic phosphate production by enzymes such as phosphatases and ATPases. Red spheres are oxygen, blue spheres are nitrogen, large gray spheres are carbon, small gray spheres are hydrogen, magenta spheres are phosphorus, and cyan spheres are molybdenum.

of *S. cerevisiae* Cdc14 (ScCdc14). This is a very important comparison. If FoCdc14 substrate specificity is essentially identical to that of ScCdc14, then the active sites of these two enzymes must be essentially the same. This has two important consequences. First, it implies that the existing crystal structure of ScCdc14 bound to a substrate peptide is likely an accurate representation of how FoCdc14 binds substrates (we don't have structural information on substrate binding by the FoCdc14 active site). Second, it suggests that inhibitors that were previously identified to work against ScCdc14 or were designed based on the ScCdc14 structure are likely to work against FoCdc14 (and by extension Cdc14 from many other fungal pathogen species that have similar substrate specificity). In contrast, if FoCdc14 specificity is very different, it would argue that Cdc14 active sites vary significantly by species and that inhibitors that bind the active site of one Cdc14 probably won't bind other orthologs well. In this case, Cdc14 wouldn't be a good broad-spectrum antifungal target.

This week's lab will require you to revisit the Cdc14 literature and find what is known about ScCdc14 specificity. It is well characterized in several papers that use the same kind of synthetic phosphopeptide substrates and enzyme assay you will use in today's lab. This literature will allow you to predict how FoCdc14 should act on the collection of phosphopeptides if the hypothesis that its active site is the same as ScCdc14 is correct.

LAB INSTRUCTIONS

PART 1. LITERATURE ANALYSIS

The specificity of Cdc14 phosphatases has been characterized to some extent. You have been provided with four literature articles (PDF files) that document substrate specificity features of Cdc14 phosphatases.[1] Scan through the relevant part of the results section and figures in each to learn what features of a substrate Cdc14 recognizes and requires for efficient catalysis.

In lab today you will be comparing FoCdc14 activity toward a panel of similar but distinct phosphopeptide substrates that mimic the biological substrates that Cdc14s would encounter. The phosphopeptide sequences you will characterize are listed in Table 11.1 using single-letter amino acid code. Note that phosphorylated

TABLE 11.1 *Phosphopeptide Substrate Sequences*

#1	HT{pS}PIKSIG
#2	HT{pT}PIKSIG
#3	HT{pY}PIKSIG
#4	HT{pS}AIKSIG
#5	HT{pS}PIASIG
#6	HT{pS}PIRSIG
#7	HT{pS}PKASIG
#8	HT{pS}PIAKIG
#9	RT{pS}PIKSIG
#10	HD{pS}PIKSIG
#11	HT{pS}PIKRIG
#12	HT{pS}PIKKRG

amino acids are denoted with a lowercase p in curly brackets (e.g., {pS} for phosphoserine). Sequence #1 is the natural sequence from a known ScCdc14 substrate, the DNA repair enzyme Yen1.[2] The other peptides have individual positions changed to another amino acid.

Based on your reading of the provided articles, generate a hypothesis for which substrates should be the best and which should be the worst. Group them into four hypothetical categories: excellent, good, average, and poor substrates. Make sure you are able to justify your hypothesis with evidence from the literature. You will then be estimating k_{cat}/K_m catalytic efficiency values for each substrate as discussed above in the "Background and Theory" section under identical reaction conditions to test your hypothesis.

PART 2. CATALYTIC EFFICIENCY MEASUREMENTS

1. **Standard curve.** Yes, you need another standard curve to convert absorbance measurements to product formed.
   A. You have been given a series of sodium phosphate standards (0, 2.5, 5, 10, 20, 30, 40, and 50 μM) in reaction buffer.
   B. Transfer 50 μl of each into the first column of your microplate.
   C. Add 100 μl malachite green reagent to each, mix carefully to avoid bubbles, and incubate for 20 minutes to allow the color to fully develop.
   D. Measure the absorbance at 640 nm using the plate readers, export the data as an Excel file, and transfer to your computers.
2. **Make your enzyme stock.** Make a working 10x enzyme stock at a concentration of 1 μM by diluting your main FoCdc14 enzyme stock in reaction buffer. Make at least 250 μl of this solution. Some groups may need to adjust this target concentration based on the results of the k_{cat} analysis last week. Consult with the instructor first before making your stock.
3. **Make reaction master mixes.** You will be performing your reactions for the 12 phosphopeptide substates and a blank in triplicate (13 reactions per trial, 39 total reactions). Make a master mix of the common reagents (buffer and enzyme) sufficient for 14 reactions for each trial:
   A. 40 μl reaction buffer x 14 = 560 μl
   B. 5 μl of the 10x FoCdc14 solution (from step 2) x 14 = 70 μl
   C. Mix the buffer and enzyme together thoroughly in a microfuge tube.
   D. Aliquot 45 μl into each reaction well. Set your plate up so that columns 2 and 3 have trial 1 reactions for the 1–6 and 7–12 phosphopeptides, respectively (rows A–F), and the blank can go in 2G. You will then repeat this with columns 4–5 and 6–7 for the other two trials.
4. **Load plate wells with substrate.** Transfer 25 μl of substrates 1–6 into rows A–F of column 8 and 25 μl of substrates 7–12 into rows A–F of column 9. Add 100 μl of reaction buffer to well 8G.
5. **Start reactions.** Use the multichannel micropipettor to transfer 5 μl of substrate into the reaction wells to start the reactions. Do this in the following order. With an initial set of tips, draw up 5 μl substrate from wells 8A–8G and mix in reaction wells 2A–2G, mixing carefully by pipetting up and down a few times to the first stop position. Without changing tips, repeat for columns 4 and 6. Change to a fresh set of tips and draw up 5 μl from wells 9A–9F and dispense into 3A–3F, mixing carefully. Without changing tips, repeat for columns 5 and 7. Do this carefully but quickly so that the reactions all start around the same time.
6. Allow reactions to proceed for 20 minutes at your bench. While the reactions are proceeding, carefully load 325 μl malachite green dye solution into wells A–G of column 10 and A–F of column 11.

7. **Stop reactions.** At the 20-minute mark, use the multichannel micropipettor to transfer 100 μl of malachite green dye solution to each reaction well. Do it in the exact same order that you started the reactions to ensure that the reaction times are as close to identical as possible. Set the multichannel pipettor on 50 μl (its maximum volume). You will need to transfer 50 μl twice for each column of reactions. Transfer the 100 μl to all three sets of reaction wells first, then go back and mix each set by carefully pipetting up a down a few times. The dye is acidic and will denature the enzyme, effectively stopping the reactions.
8. **Measure absorbances.** Wait for 20 minutes after adding the final aliquots of malachite green dye. Insert the plate into the plate reader and measure the absorbance values at 640 nm for all reaction and blank wells (columns 2–7). Export the absorbance values as a new Excel file and transfer to your lab computers.

DATA ANALYSIS AND SUBMISSION

Use official Excel functions for all calculation steps. Remember, once you write a function in one cell, you can drag the box in the lower right corner to copy the equation into adjacent rows and columns. This makes it very quick to apply the same calculation to a bunch of cells. Also, remember to use the "$" in front of the column and row coordinates for fixed values to prevent them from changing when the equation is copied to other cells.

1. You have been provided with an Excel spreadsheet template for data analysis in your group data storage folders. It is titled "Week 11 Data Analysis Template." Open it in Excel on your lab computer.
2. Paste the raw absorbance values for your three reaction trials from your exported data file into the appropriate cells (B3:D15) at the top left of the spreadsheet. This includes your three blank absorbance values in B15:D15. Calculate the average blank absorbance in cell E15.
3. Paste the raw absorbance values obtained for your phosphate standards from the exported data file into cells K3:K10 in the analysis template. Generate a standard curve from the absorbances of your standards. Be sure to subtract the 0 μM phosphate standard (blank) from the other standards first (results should go in L3:L10). Your standard data will automatically be graphed. Perform linear regression and force your fit line through the origin so there is no y-intercept. Check the box to include the regression equation on the graph. Type your regression slope into the green highlighted cell above the graph (M12).
4. Type the total reaction time (in seconds) into cell E31. Type your final enzyme concentration (in μM) into cell B45. Type the final substrate concentration (in M) into cell B46.
5. The next step is background subtraction. Subtract the average background (cell E15) from all 36 reaction absorbances in the Blank Correction section (cells B18:D29).
6. Convert the corrected absorbance values from step 5 to reaction velocities (μM phosphate/sec) in B32:D43 using the standard curve slope (cell M12) and reaction time (cell E31).
7. Calculate k_{cat}/K_m for all reactions in B48–D59. In this calculation you must divide the reaction velocities from step 6 by the enzyme concentration (B45) to get the k_{cat} units of sec^{-1}. As long as the product and enzyme units are the same (μM in this case), they will cancel out. Then, you also must divide by the substrate concentration (in M) to get the K_m units of M concentration in the denominator. The final result is the k_{cat}/K_m in units of sec^{-1} M^{-1}.
8. Calculate the average k_{cat}/K_m and standard deviation of each of the substrates in F48:G59.
9. Generate a bar graph comparing the average k_{cat}/K_m from all phosphopeptides (with standard deviations as error bars). You will need to use this graph as a figure in your final report.

10. Student's t-test is a statistical procedure to compare two average values to determine if they are significantly different. In Excel, use the t-test function to determine if the average catalytic efficiency for the wild-type peptide sequence (#1) is statistically different from each of the other peptides (this will require doing separate t-tests for each pair of peptide k_{cat}/K_m values). Consider p values <0.05 as significant. This may identify both positive and negative determinants of activity (changes that increase or decrease activity).
    A. In cell H49 access the T.TEST function from the Formulas menu → More Functions → Statistical. Select the three k_{cat}/K_m values for peptide #1 as Array1 and the three values for peptide #2 as Array2. Enter "2" in the Tails box and "3" in the Type box.
    B. Click OK to perform the test. The calculated p-value will appear in the cell.
    C. Click on the cell to see the T.Test parameters. Fix the Array 1 values by inserting "$"—it should look like $B$48:$D$48. Then you can click on the box in the lower right corner of the cell and drag it to copy the t-test to cells H50:59. You will now have a p-value indicating for each peptide whether its catalytic efficiency is significantly different than peptide 1.
11. Enter your answers to the queries in rows 62–65.

Save your analysis Excel file to your group data storage folder and submit it to the course learning management system before leaving lab today. Make sure you have stated your conclusion about how similar the FoCdc14 specificity is to Cdc14 from *Saccharomyces cerevisiae* in row 65. This last part is essential for validating your use of the ScCdc14 structure as a model for FoCdc14 to design inhibitors and must be included in the final reports.

12. ENZYME INHIBITION

BACKGROUND AND THEORY

ENZYME INHIBITORS

Enzyme inhibition occurs as a natural regulatory mechanism in biological systems. It is an easy way to control the output of metabolic pathways. Enzyme inhibition is also relevant to human health, as many therapeutic drugs are enzyme inhibitors. Inhibitors can affect enzymes in a variety of ways. They can reduce substrate binding, for example through competition. They can also interfere with the chemical reaction steps. Some inhibitors bind reversibly to their enzyme targets, whereas others irreversibly bind their targets through covalent bonds.

There are three types of reversible enzyme inhibitors: competitive, noncompetitive (including mixed), and uncompetitive (Figure 12.1).

- Most competitive inhibitors are structurally similar to the enzyme's substrate and bind the enzyme active site, effectively blocking substrate binding. Inhibition can be overcome though at high substrate concentrations.
- Noncompetitive inhibitors interact with the enzyme somewhere other than the active site and cannot be overcome by high substrate concentrations. These are often called allosteric inhibitors. Mixed inhibition is a form of noncompetitive, or allosteric, inhibition where binding of the inhibitor at the allosteric site also partially affects binding of substrate at the active site. This can occur when allosteric binding alters the overall structural conformation of the enzyme, including the active site.
- Uncompetitive inhibitors interact only with the enzyme-substrate complex. This is a relatively rare type of inhibition, and like noncompetitive, it cannot be overcome by increasing the substrate concentration.

CHARACTERIZING REVERSIBLE INHIBITORS

The three types of reversible enzyme inhibitors have unique effects on the steady-state kinetic parameters V_{max} (or k_{cat}) and K_m (Figure 12.2). This means that you can characterize the mechanism of a reversible inhibitor by conducting the same steady-state assays you performed in Week 10 in the presence of varying concentrations of inhibitor.

- Competitive inhibitors cause an increase in the measured K_m (apparent decrease in substrate binding affinity) but do not affect V_{max}.
- Noncompetitive inhibitors cause a decrease in V_{max} (slower catalytic turnover rate) but do not alter the measured K_m.
- Note that mixed inhibition causes both an increase in measured K_m and a decrease in measured V_{max}.
- Uncompetitive inhibitors cause a decrease in both K_m and V_{max} (apparent increase in substrate binding affinity with a corresponding decrease in catalytic turnover rate).

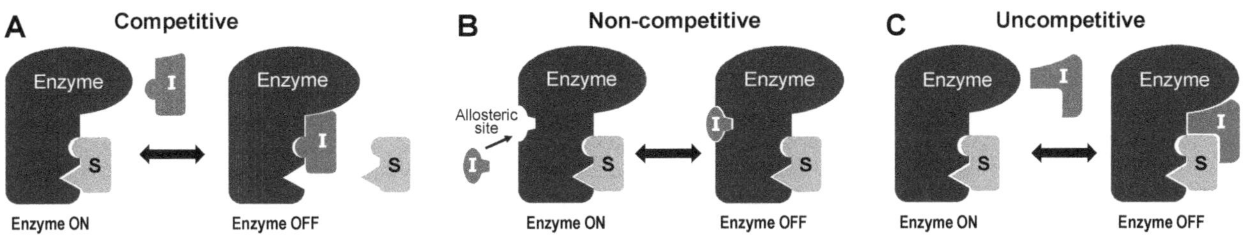

FIGURE 12.1. (A) Competitive inhibitors compete with substrates for binding to the enzyme's active site. This lowers activity by reducing the rate at which enzyme-substrate (ES) complex is formed. (B) Noncompetitive inhibitors, or allosteric inhibitors, interact with the enzyme at a site distinct from the active site and do not influence ES formation. However, they reduce the rate at which substrate is converted to product. Mixed inhibitors are allosteric inhibitors that also reduce ES complex formation, for example by altering the conformation of the active site. (C) Uncompetitive inhibitors only bind to the ES complex.

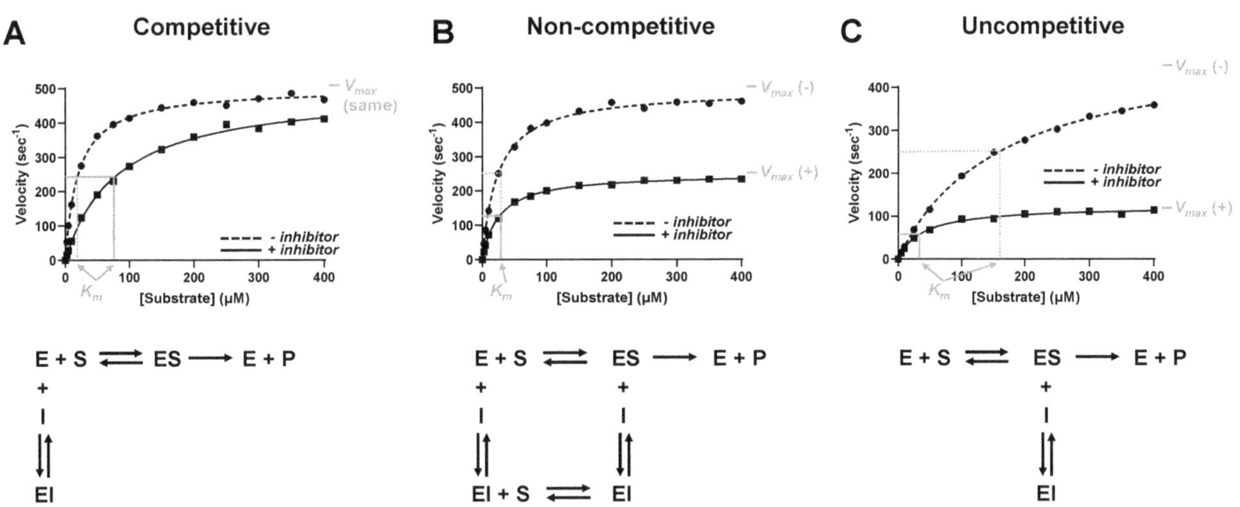

FIGURE 12.2. The different types of reversible enzyme inhibitors have unique effects on the steady-state kinetic parameters k_{cat} (V_{max}) and K_m, allowing the mechanism of inhibition to be determined from the same velocity versus substrate concentration experiments covered in Week 10, conducted with varying inhibitor concentrations. (A) Competitive inhibitors do not alter V_{max} and increase the apparent K_m. Inhibition can be fully overcome with enough substrate, since substrate and inhibitor compete for binding to the active site. (B) Noncompetitive allosteric inhibitors reduce the V_{max} but do not alter the K_m. (C) Uncompetitive inhibitors lower both the V_{max} and the apparent K_m. Mixed allosteric inhibitors will lower the V_{max} and increase the apparent K_m (not shown).

See the Week 12 lecture for explanations of why the different inhibitor mechanisms affect the steady kinetic parameters in unique ways.

K_i is a special term called the inhibition constant that represents the affinity of the inhibitor for the enzyme. It is a fundamental property of an enzyme inhibitor, much like K_m is a fundamental property of an enzyme substrate. K_i is an equilibrium constant defined as the ratio of k'_{-1}/k'_1, or [E][I]/[EI] in the reaction scheme in Figure 12.3 (see this week's lecture for additional detail).

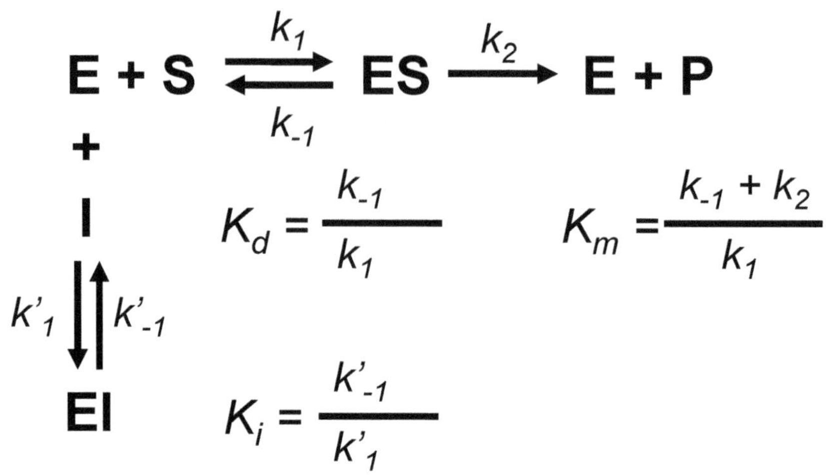

FIGURE 12.3. The inhibition constant, K_i, and the substrate dissociation constant, K_d, represent the affinities of an inhibitor and a substrate for an enzyme, respectively. They are both equilibrium constants, defined as the concentration of the ligand (inhibitor or substrate) required for half of the enzyme molecules to be in the bound state at equilibrium.

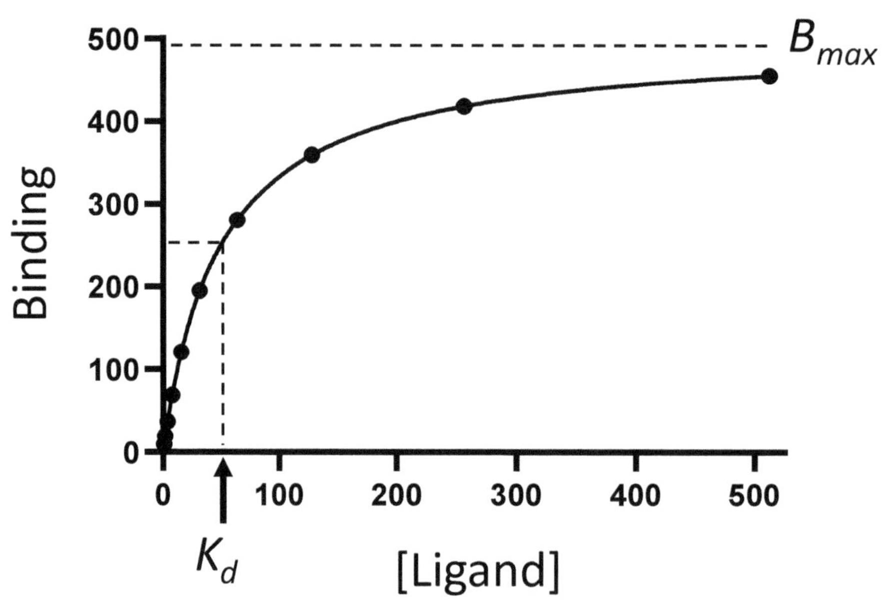

FIGURE 12.4. Binding of a ligand, such as an enzyme inhibitor, to a protein can be measured with a variety of assays. Measuring binding as a function of ligand concentration results in the same hyperbolic curve observed with steady-state enzyme kinetics, also due to the saturation effect. In this case, the asymptote is B_{max} (maximum binding), representing full occupation of all binding sites by ligand. K_d (or K_i in the special case of an enzyme inhibitor) represents the ligand concentration that results in half of the maximum binding, just like K_m was the substrate concentration resulting in half maximal reaction velocity.

Functionally, K_i is the concentration of a reversible inhibitor at which half of the enzyme molecules are bound and is identical to the dissociation constant, K_d, for any ligand binding to a protein (Figure 12.4). Measuring K_i is a useful way to compare different inhibitors of an enzyme (you can imagine how this would be useful in drug discovery, providing a way to sort and prioritize the most potent inhibitor compounds for a target enzyme of interest). K_i can be calculated from steady-state kinetics experiments run at different inhibitor concentrations such as those illustrated in Figure 12.2. However, this is outside the scope of your projects in this course.

CHARACTERIZING IRREVERSIBLE INHIBITORS

Many powerful drugs act by irreversibly binding their targets through a covalent bond. Aspirin and penicillin are two prominent examples of irreversibly acting drugs. Determining if a newly identified inhibitor acts reversibly or irreversibly is an important early step in its characterization. For example, irreversible active site inhibitors actually behave like noncompetitive inhibitors in steady-state kinetic experiments under some conditions even though they bind the same site as the substrate. This can confuse mechanistic interpretations if it is not recognized that they are binding irreversibly.

There are a variety of methods to detect irreversible inhibition, but one of the simplest is measuring the effect of time on inhibition. Under steady-state conditions, reversible inhibitors will rapidly establish a binding equilibrium with the enzyme, after which a constant initial velocity will be observed (linear increase in product, as you learned in Chapters 9 and 10). However, many irreversible inhibitors will gradually form covalent bonds with the enzyme, permanently reducing the pool of active enzyme and thereby continually lowering the apparent reaction rate. This nonlinear behavior is easily detectable by monitoring the reaction rate over time under otherwise steady-state conditions (Figure 12.5). You will look for this signature today in the candidate Cdc14 inhibitors you will work with.

INHIBITOR SCREENS

In the world of drug discovery, a common experiment is to screen large libraries of chemical compounds for a desirable property, such as inhibition of an enzyme linked to a disease. To identify the best inhibitors and candidate drug leads, researchers must design their enzyme assay very carefully and validate it first with controls. For example, when looking for competitive active site inhibitors, competition with substrate is a factor that must be considered. If a saturating concentration of substrate is used, the effects of inhibitors might be obscured. That is why knowing the K_m for the substrate becomes important. A substrate concentration can be chosen that ensures sufficient measurable activity but will not completely outcompete inhibitor molecules for active site binding. In order to identify the best inhibitors, sufficient dynamic range is needed for detection of statistically significant differences between inhibitors. This requires that activity in the absence of inhibitor be near the upper end of the dynamic range. Knowing the k_{cat} and K_m for the substrate and the dynamic range of the assay (e.g., from the standard curve) allows the researcher to easily establish conditions to ensure sufficient initial activity. Researchers typically run a quality control test on their screen assay, called a Z′ analysis, to determine if the design is suitable for reliably detecting inhibition. It will be your responsibility today to design and perform a suitable active site inhibitor screen for FoCdc14 using pNPP as the substrate.

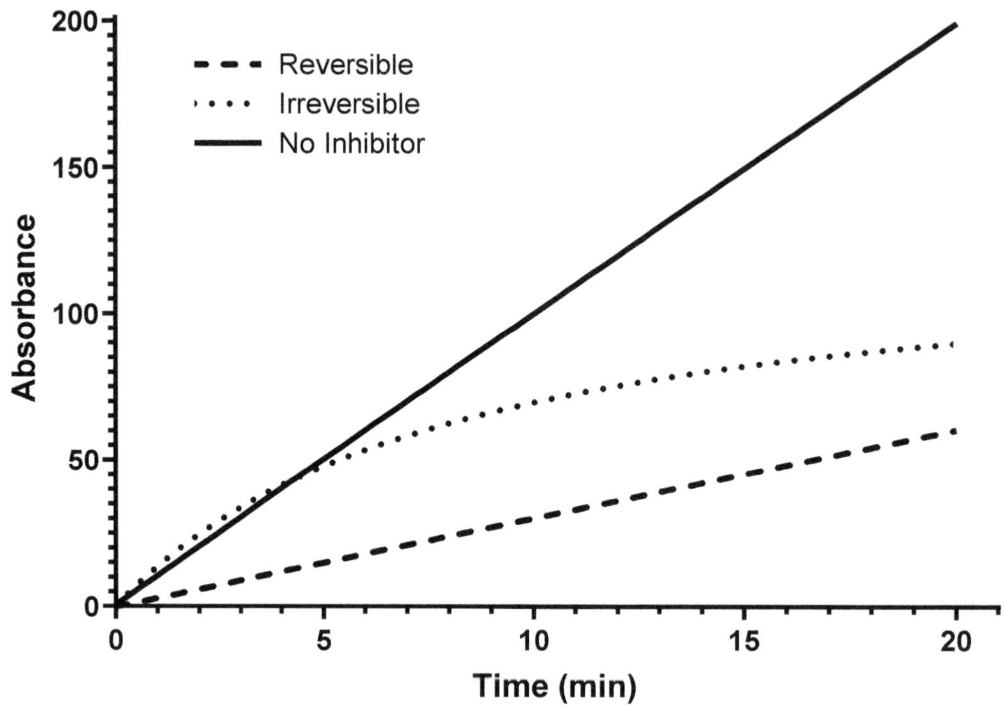

FIGURE 12.5. Under steady-state conditions, formation of product remains constant over time. A typical reversible competitive inhibitor reduces the rate of product formation (decrease in slope of the linear progress curve) consistently throughout the reaction. A typical irreversible competitive inhibitor gradually reduces the amount of enzyme available to catalyze the reaction by permanent inactivation. This results in a gradual decrease in slope over time and nonlinear progress curve. This type of progress curve can be useful in revealing if an inhibitor is acting irreversibly.

OBJECTIVES

The primary objective is to identify irreversible inhibitors of FoCdc14. Cdc14 enzymes have a highly reactive cysteine in the active site—the amino acid you mutated earlier in the semester. This reactive Cys is prone to irreversible inhibition by electrophilic molecules. You specifically want active site inhibitors so you can use the known active site structure from *S. cerevisiae* Cdc14 (RCSB PDB ID 5XW5) to design optimized inhibitors. Because you don't have time to characterize the mechanism of candidate inhibitors using steady-state kinetics, you can have some confidence that they are binding the active site if you find ones that act irreversibly. As noted above, under steady-state conditions an irreversible inhibitor will exhibit time-dependent inhibition. The longer the reaction, the slower the enzyme rate as more and more enzyme molecules become inactivated. In contrast, inhibition by reversible inhibitors is generally independent of time during the steady state. Today you will compare inhibition of FoCdc14 by several small molecule compounds previously identified as inhibitors of *S. cerevisiae* Cdc14 in a high throughput chemical library screen. By measuring inhibition continuously using the pNPP assay, you will be able to detect ones that act irreversibly on FoCdc14 and select the most effective one as your lead compound for computational docking and optimization next week.

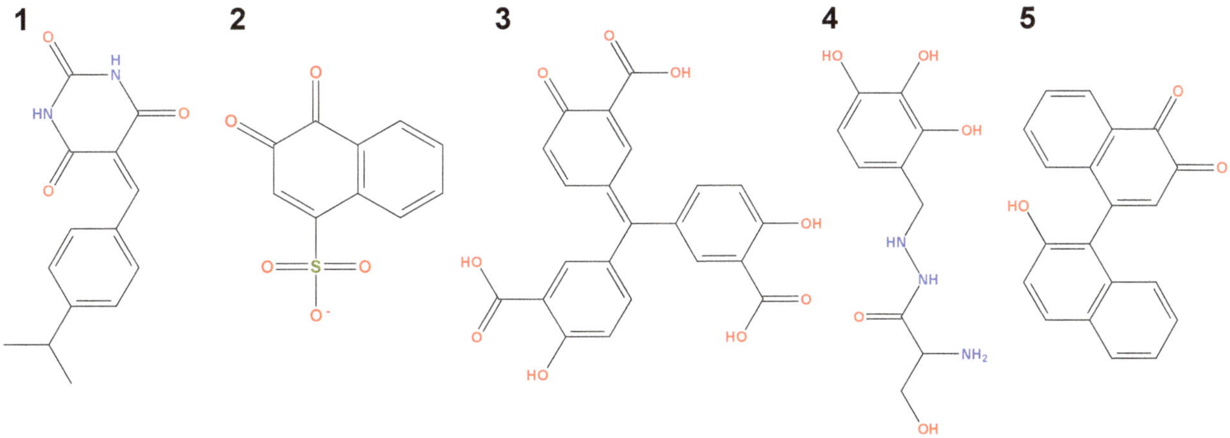

FIGURE 12.6. The chemical structures and names of the five candidate inhibitors that will be compared in the screening experiment this week. All were previously identified as inhibitors of the *Saccharomyces cerevisiae* Cdc14 phosphatase in chemical library screens. 1: 5-(4-isopropylbenzylidene)-2,4,6(1H,3H,5H)-pyrimidinetrione; 2: 3,4-dioxo-3,4-dihydro-1-naphthalenesulfonate; 3: Aurintricarboxylic acid; 4: DL-Serine 2-(2,3,4-trihydroxybenzyl)hydrazide; and 5: 2′-hydroxy-1,1′-binaphthalene-3,4-dione.

To accomplish the objective, your group will need to design an assay to screen FoCdc14 for inhibition by these candidate inhibitors. You will use the pNPP assay from previous weeks, but you will have to define the appropriate conditions to measure inhibition by these compounds. This includes suitable enzyme and substrate concentration. You will run the reactions for 30 minutes to provide enough time to observe evidence of irreversible inhibition. Remember that you want active site inhibitors, so competition with substrate will be a factor in your reactions. Your previous kinetics experiments have the information you need to design this assay. The importance of knowing the k_{cat} and K_m becomes clear when you have to design experiments such as this. You will need to perform enough independent trials to assess reproducibility and statistical significance of differences in inhibition between these compounds. Figure 12.6 shows the structures of the candidate inhibitors you will compare.

LAB INSTRUCTIONS

1. Create a reaction table that lists all of the ingredients of each reaction you will run with stock concentrations as the column headers and volumes used in the individual cells (similar to tables you have used in previous weeks). For this experiment, the inhibitors will be fixed at 100 μM final concentration. Reaction time is fixed at 30 minutes. Use your standard curve and k_{cat} and K_m values to determine appropriate enzyme and substrate concentrations to get high level activity in the absence of inhibitor. This provides a broad signal range in which to see varying levels of inhibition. The chosen substrate concentration should avoid enzyme saturation.
2. Make any calculations needed to make your stock solutions (e.g., enzyme, substrate, and inhibitor solutions). You will be provided with a 500 mM pNPP solution, your FoCdc14 enzyme, reaction buffer, and a 20 mM stock of each inhibitor dissolved in the solvent dimethyl sulfoxide.
3. Set up your reactions so that you are adding enzyme to start them (that way the enzyme sees the substrate and inhibitor at the same time). Monitor product formation over time in the plate reader like you did in Week 10 when performing

steady-state kinetics. Look for two things: (1) a reduction in the rate of product formation (absorbance) in the presence of inhibitor compared to the absence of inhibitor and (2) a decrease in the slope over time (nonlinear behavior) consistent with irreversible enzyme inactivation (see Figure 12.5).

4. Of the inhibitors demonstrating irreversible behavior, calculate % inhibition at the final timepoint. Apply statistical tools you learned in previous weeks to assess reproducibility and significance of any differences between inhibitors.
5. Select your best inhibitor based on (1) evidence of irreversible behavior and (2) highest % inhibition.

DATA ANALYSIS AND SUBMISSION

Save a single Excel file to your group data storage folder and submit to the course learning management system that includes your experimental setup, raw data, all calculations, and graphs of product formation over time. Also include a statement of which inhibitor you have chosen to move forward with as your lead compound next week along with an explanation of why.

13. PROTEIN STRUCTURE MODELING

BACKGROUND AND THEORY

Computational modeling tools can be an efficient and valuable approach to studying the interaction of small molecules with proteins that have well-defined atomic resolution structures. Applications include the identification and design of inhibitors in the drug discovery field as well as mechanistic understanding of how enzymes recognize and act on substrates.

DETERMINING PROTEIN STRUCTURE

Prior to 2021, reliable atomic resolution protein structures had to be determined in the lab using sophisticated and laborious methods. Interpreting diffraction patterns obtained from protein crystals irradiated with an X-ray beam (often called X-ray crystallography) was the first method developed and is still a common state-of-the-art approach. However, it has several challenges. First, it requires relatively large amounts of highly purified protein under specific solution conditions that will allow crystals to grow. This is challenging or impossible for many proteins and protein complexes. Second, it requires access to powerful X-ray sources that are usually only found at specialized facilities, called synchrotrons (circular particle accelerators), such as the Advanced Photon Source at Argonne National Lab.

A more recent method is cryo-electron microscopy (cryo-EM). This method requires much less material, can tolerate impurities, and is applied to liquid solutions (no crystals needed) under physiological conditions. In cryo-EM, protein samples are frozen in a very thin layer on a metal grid in a way that prevents ice crystals from forming (called vitrification). The samples are then subjected at different angles to a beam of electrons and the "shadows" cast by protein molecules are recorded on a detector. Computational tools are used to generate a 3D model of the proteins from the collected images. See the lecture material for additional information and resources on these methods for protein structure determination.

In 2021 scientists at Google DeepMind solved the protein folding prediction problem, described in a landmark 2021 *Nature* paper, with a deep learning algorithm they called AlphaFold2.[1] The developers were awarded the Nobel Prize in Chemistry in 2024. For the first time, it became possible to generate reliable atomic resolution protein structures computationally. The AlphaFold2 algorithm is based on evaluating large multisequence alignments of orthologs and assessing the evolutionary changes that occur in the amino acid sequence. The algorithm uses this information to predict which amino acids must be close to each other in space. It then uses known physicochemical properties of amino acids and the atomic bonds present in proteins to construct and iteratively refine a realistic 3D model. It has now been well established that AlphaFold2 models can achieve similar accuracy to experimental models obtained by X-ray crystallography and cryo-EM for many individual proteins. Improvements in predicting oligomeric protein complexes and complexes between proteins and ligands have been made with the AlphaFold3 algorithm.

COMPUTATIONAL DOCKING

With high-resolution structure models of a protein and a ligand and knowledge of the general location of the binding site (such as an enzyme active site), computer algorithms can predict how the ligand will interact with the protein. They can even predict the affinity of binding. This process is called docking. Docking can be used as a tool to virtually screen large libraries of chemical structures for molecules predicted to bind a target of interest with high affinity. Virtual screens are fast, inexpensive, and unlimited by practical considerations such as having actual molecules synthesized and purified. Once hits are identified, a relatively few molecules can be synthesized and tested in the lab for target binding and other properties. Once a lead compound has been identified in the drug discovery process, docking is also useful as a starting point for designing further improvements in the lead's structure to enhance affinity, specificity, and drug-like properties (solubility, stability, ability to cross biological membranes, etc.).

The basic steps in computational ligand docking are the following. First, the receptor protein and the ligands must be prepared to represent physiologically accurate molecules. This includes adding hydrogen atoms (which are typically absent from high-resolution structures) and adding appropriate charge states to functional groups at physiological pH. Next, the boundaries for the docking area need to be defined. It is computationally challenging to allow the ligand to dock anywhere on the protein. It typically works best when a specific binding pocket is defined, such as the active site of an enzyme or a known ligand binding site on the protein. Finally, the computer samples various orientations of the ligand in the docking area, calculating interactions between functional groups on the ligand and the receptor for each pose to come up with a set of preferred poses and docking scores, which are usually an estimation of the binding free energy, ΔG. The scores are based on known properties of molecular bonds and the chemical properties of proteins and small molecules. It is common to allow flexibility of the ligand during the docking process, such as free rotation of single bonds. The protein receptor can be kept as a static, rigid structure or can be allowed to have limited flexibility.

OBJECTIVES

In today's lab you will first generate a 3D model of FoCdc14 using the newest AlphaFold algorithm, AlphaFold3. You will compare your structure model to the experimentally determined structure of *S. cerevisiae* Cdc14 (ScCdc14). You will then perform a docking experiment to predict how the inhibitor you selected last week binds in the enzyme active site. Finally, you will evaluate your docked structure to design an improvement to your inhibitor based on the observed properties of the Cdc14 active site. It would be ideal to use substrate binding as a guide for the inhibitor design, since many enzyme inhibitors and drugs act by mimicking substrate structure. However, due to time constraints you will undertake a simpler design task.

The identification of inhibitors through methods such as high-throughput screening of chemical libraries is usually only the first step in development of a useful drug (or pesticide). Additional rounds of modification and optimization of various properties of a compound are typically necessary to create something that is maximally effective, cost-effective to produce, stable, safe, etc. Your ligand docking and optimization exercises today should give you a brief glimpse of what this process is like once a lead compound has been identified for a given target enzyme. The new compounds you design would then need to be synthesized and tested for improved performance (e.g., lower K_i values, improved specificity for inhibiting Cdc14 compared to other phosphatases, activity toward live fungal cells, etc.) if the project were to continue.

LAB INSTRUCTIONS

PART 1. GENERATE A PREDICTED STRUCTURE OF FoCdc14 USING ALPHAFOLD

1. Access the AlphaFold server at https://alphafoldserver.com/. Note that this requires registering with a Google account but is completely free to use.
2. Retrieve your FoCdc14 translated amino acid sequence (from Week 1) and paste it into the query window, making sure that Entity type is set to protein and Copies is set to 1 (if you want you can try for fun later setting Copies to 2, since you now know that Cdc14 proteins exist as stable dimers).
3. When the prediction is complete, a check will appear next to the job name. Click the 3 vertical dots at the right side of the job line and download the results. Extract all files from the .zip folder. The only files you need are the .cif structure model files. By default, AlphaFold generates five models. We only need to use one, but save the five .cif structure files to your group's data storage folder.
4. Retrieve the .cif file for the published crystal structure of *S. cerevisiae* Cdc14 bound to an optimal phosphopeptide substrate from your Week 1 data storage folder and copy it to your Week 13 folder. In the event you are unable to locate it, go to the Protein Data Bank (https://www.rcsb.org/) and search for "5XW5." On the 5XW5 main page, open the Download Files dropdown menu and select "PDBx/mmCIF Format." Transfer the 5XW5.cif file to the Week 13 folder.

PART 2. COMPARE YOUR ALPHAFOLD MODEL TO ScCdc14 STRUCTURE

1. Open PyMOL on your computer. PyMOL is a commonly used molecular visualization tool from Schrodinger LLC that is free for educational purposes.[2] The graphical user interface for PyMOL is shown in Figure 13.1, with key areas and features labeled as they are referenced below.
2. Load the 5XW5 structure using "File" → "Open" from the main menu area and navigating to the file storage location. The structure is a dimer, and only one subunit has the phosphopeptide substrate bound. By default, you will see the secondary structure display. This is fine for our purposes now. You need to delete the other subunit and solvent molecules for simplicity. Under the Display menu, click "Sequence." You should see a line of amino acid sequence appear at the top of the viewing window. Under "Display," click "Sequence Mode" → "Chain Identifiers." Now the line should show just the letter codes of the various chains in the structure. Left-click on the second chain (designated with a green "B") to select it. Then right-click on it and select "remove." The second chain should disappear. Do the same thing for the last four chains (red letters B, A, B, and C), selecting "remove" after right-clicking. You should now have two chains remaining, A (one ScCdc14 subunit) and C (the phosphopeptide substrate bound to this subunit's active site).

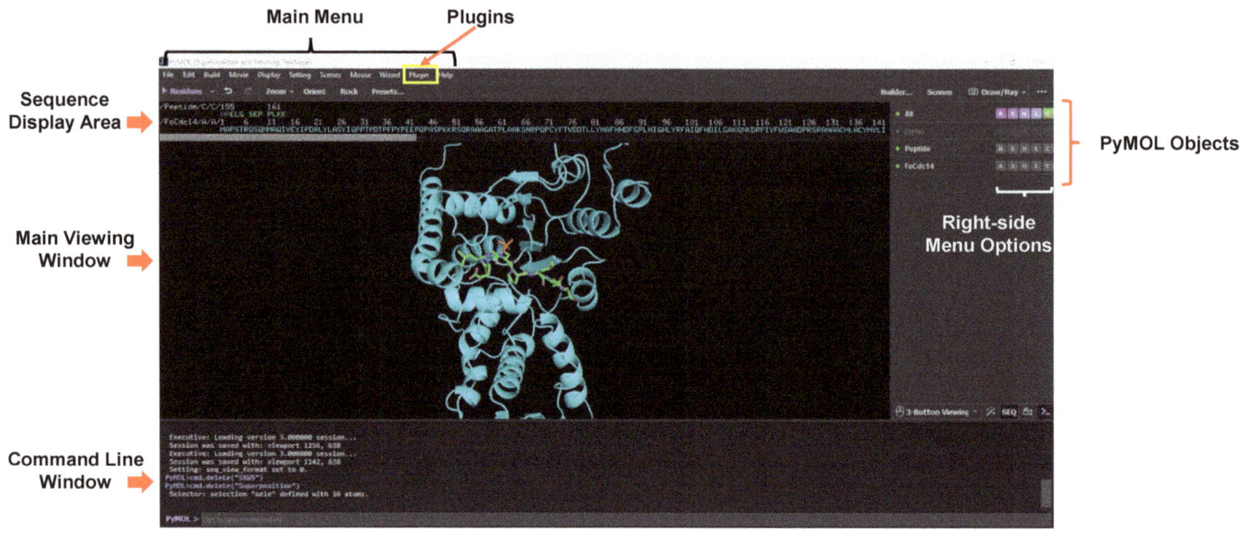

FIGURE 13.1. A screenshot of the graphical user interface of the PyMOL Molecular Graphics System software, version 3.1.6.1 (Schrödinger, LLC; RRID: SCR_000305). Several key work and menu areas needed for completing the lab experiments described in this chapter are labeled for guidance.

A. Click on the A chain to select it. On the right-side menu, click the "A" button on the (sele) line and click "center" to center the view window on the selected object. Click in an empty black area of the viewing window to deselect the A chain.

B. Click on the C chain (peptide substrate) to select it. In the right-side menu click the "A" button on the (sele) line, then click "extract object." This will move the phosphopeptide to its own structure object separated from the 5XW5 protein chain.

C. In the right-side menu select "A" in the obj01 line and choose "rename object." Change the name from "obj01" to "Peptide."

3. Next, open one of the AlphaFold models of FoCdc14 (.cif files). It shouldn't matter which one; they are likely to all be very similar. You can rename the chain in the right-side menu using "A" → "rename object." Call it "FoCdc14." With both structures now open, go to the Plugin menu and select the "Alignment" tool.

A. Click the "One to one" radio button. Make sure "align" is selected in the "Method" dropdown menu. In the "Mobile Selection" window choose the FoCdc14 structure. In the "Target Selection" window choose the 5XW5 structure. In "Create alignment object" enter the name "Superposition."

B. Leave other settings as default and click "OK" at the bottom. Close the Alignment window. You should see that the FoCdc14 structure has moved and is overlaid on the 5XW5 structure. What PyMOL has done here is create a sequence alignment (much like Clustal omega does) and then superpose the two structures based on this alignment. This is a great way to visually see how similar the AlphaFold structure prediction of FoCdc14 is to the solved structure of ScCdc14.

C. In the right-side menu click "H" on the Superposition line and then "cgo" to remove all the little yellow alignment dashes.

D. Look in the lower PyMOL Command Line window. You should see some text output from the Alignment results. Look for the line that reads "Executive: RMSD =." It is the calculation of the overall RMSD for the superposed structures. RMSD stands for "root mean square deviation." It reflects the average distance between aligned atoms in the two structures. RMSD values < 1 Å are expected for proteins that are almost identical. RMSD values between 1 and 2 Å are typical for homologous proteins with very similar structures. RMSD values of 2–3 Å would be interpreted as similar folds but with noticeable differences. RMSD values > 3 Å would indicate poor similarity between the structures. Record your RMSD value. How similar is your FoCdc14 model to the ScCdc14 structure?

E. In the right-side menu click "A" on the (sele) line and select "center." This recenters the superposed structure for easier manipulation and viewing. Note that you can zoom in and out by right-clicking and dragging up and down on the viewing window. You can also slide the structure around in the plane of the screen by holding Ctrl and left-clicking and dragging and can control the depth of the view by using the mouse wheel. Finally, you can rotate the structures about the center by clicking and holding the left mouse button and dragging.

4. Go back to the Display menu and select "Sequence Mode" → "Residue Codes." Now you will see the amino acid sequence for all three chains with the two protein chains aligned (just like in the Clustal omega tool you used earlier). Scroll to the end of the sequences. You will notice that FoCdc14 has extra sequence at the end not present in 5XW5. Click and drag on all the sequence at the end of FoCdc14 that does not have corresponding 5XW5 sequence. Right-click on this selected sequence and choose "remove." You should see the selected sequence disappear from the sequence display and from the structure itself.

A. Hide the Peptide chain by clicking on its line in the right-side menu.

B. Rotate the superposed structure now to inspect how well or how poorly the two structures match. Take note of any regions that do not overlay nicely or do not have matching overlaid secondary structures. To identify the amino acids corresponding to regions that do not overlay nicely, hold Shift and then left-click and drag over the misaligned region in the structures. This will select the amino acids in this region and highlight them in the sequence display. Record any misaligned regions. Make sure they don't correspond to the enzyme active site, using your multisequence alignment from Week 1 as a guide.

C. You can change the color of the superposed chains if you want using the "C" buttons on the right-side menu. You can also change the background to white if desired using the "Display" → "Background" main menu option. When happy with this, save an image of your superposed structures for your final reports. Use the Draw/Ray tool in the upper right. Change DPI to 300 and click "Draw (fast)," then click "Save Image to File." Use the name "Group#_Superpose."

D. Save your PyMOL session (.pse file) at this point, using the same "Group#_Superpose" file name.

PART 3. ASSESSING SUBSTRATE BINDING TO THE FoCdc14 ACTIVE SITE

1. For the ligand docking experiments you need to reduce the length of the phosphopeptide substrate to just the core structure that makes important contacts with the Cdc14 active site. Show the peptide again in the viewing window ("S" → "sticks" in the Peptide line). Then in the sequence display, select the first four amino acids, HREL, and the last amino acid, K, in the peptide sequence. Right-click and select "remove." You will use the position of the remaining core substrate to define the area for docking of inhibitors.
2. Now, you are going to remove the structure of ScCdc14 so that you can perform the docking with your FoCdc14 structure. In the right-side menu on the 5XW5 line select "A" → "delete object." You should see the ScCdc14 structure disappear from the viewing window and the Display section at the top of the window. Save your PyMOL session again with the file name "Group#_FoCdc14_substrate."
3. To make a nice figure for your final report, change the FoCdc14 view to surface mode (right-side menu, FoCdc14 line, "S" → "surface"). Change the color to gray (right-side menu "C" → "grays" → "gray90." Center the view on the Peptide

substrate and zoom in and rotate so the position of the substrate in the FoCdc14 active site is clear and export another image as described above (File name "Group#_FoCdc14_substrate"). You will need to click "Back" first in the Draw/Ray menu to reset. Finally, hide the surface view of FoCdc14 before moving on ("H" → "surface").

PART 4. INHIBITOR DOCKING

1. First, create structure files for the five inhibitor molecules that are needed to perform computational docking. Open the Avogadro program from your desktop and the "Cdc14 Inhibitor SMILES.txt" file provided in your group data folders. For each inhibitor, perform the following steps in Avogadro: (1) "Build" → "Insert" → "SMILES," (2) paste the SMILES code text from the .txt file and click OK, (3) "Build" → "Add Hydrogens for pH" and enter 7.4 and click OK, (4) "Extensions" → "Optimize Geometry," and (5) "File" → "Save As . . ." and select .pdb file extension and use name format Inhibitor#.pdb. Save to the same folder as your FoCdc14 structure file. Use "Edit" → "Clear" to start the next inhibitor.
2. Next, the AlphaFold FoCdc14 and inhibitor structures need to be prepared for docking. First, in PyMOL use "File" → "Open" to load the five inhibitor structures by simultaneously selecting the five .pdb files you just saved. Each one should appear as an independent object in the sequence display area and the right-side menu. Next, go to the Plugin menu and open Docking Pie 1.2. This program has already been configured for you. Navigate to the "Vina" tab. Autodock Vina is the docking program you will use.
   A. Go to the "Receptors" tab. Click the "Import from PyMOL" button and select FoCdc14, then click "Import." In the "Receptors settings" window select FoCdc14 and check the "Add Hydrogens" radio button. Click on the "Generate receptor" button. Select the 01_FoCdc14 line that appears in the window at right and click on the "Set" button at the bottom. This adds hydrogens to the structure and assigns appropriate charge states to atoms at a pH of 7.
   B. Move to the "Ligands" tab and click the "Import from PyMOL" button. Select all five of the inhibitor objects and click the "Import" button. In the "Ligands settings" box choose the "All" radio button to select all five inhibitors. Select the "All" radio button under "Active Torsions" for each one, deselect "Add Hydrogens" for each one, then click the "Generate Ligand" button. All five should appear in the box at right. Select all five in this box while holding the shift key and click the "Set" button at the bottom.
   C. Back in the PyMOL window you should notice in the right-side area a new line for Vina with the prepared receptor and inhibitors as sublines.
3. Now for the tricky part. Back in the Docking Pie window, navigate to the "Grid settings" tab. Click on the "Import Objects from PyMOL" button and choose "Peptide" in the "Selection:" dropdown menu. You will see a 3D grid appear in the main PyMOL window centered on the phosphopeptide substrate and active site of FoCdc14. You task now is to adjust the dimensions and coordinates in the lower part of the "Grid settings" window so that the grid box is as small as possible but includes all essential parts of the substrate peptide (the backbone atoms and the phosphoserine and the +1 proline and +3 lysine side chains). You can rotate the image in the main PyMOL window as you do this to help finalize your settings. This box is the space that the inhibitor structures will be allowed to move around in during the docking experiment. Once you have the grid set, confirm that it is appropriate by changing the FoCdc14 protein view to surface mode (right-side menu, FoCdc14 line, "S" → "Surface.)" Make sure the grid includes all the space available in the pockets where the phosphate and the +1 proline and +3 lysine side chains sit. Once you are happy with the settings, click on the "Set in Docking Tab" button next to the grid adjustment settings in Docking Pie.
4. It's time to run the docking experiment! Navigate to the "Docking" tab in Docking Pie. Check the boxes for the FoCdc14 receptor and all five inhibitors. Set the "Poses" to 5, "Exhaustiveness" to 10, and "Energy Range" to 3. Click

the "Run Docking" button at the bottom and then the "Start" button in the popup window. When the run is completed click "Close."

5. Navigate to the "Results" tab of Docking Pie. Each inhibitor will have its own tab of data for the five poses generated during the run. Click "Save to File" at right and select "Save table to .csv file" on each tab to save the results table to your group's data storage folder. The Affinity value is the most important. In the PyMOL main window you will see that the five docking run results have been loaded as lines on the right-side menu. The results of each docking run can be shown in the active site by selecting or deselecting them using the right-side menu (simply clicking on the line toggles between visible and invisible). You can turn off the sequence display area (Display menu → "Sequence"). A "Global Frames" window appears at the bottom of the right-side section and can be used to scroll between the five poses for the visible inhibitor runs. You can view them with or without the peptide also shown. Note the consistency of the poses for each inhibitor. What does this suggest?
   A. Record which pose of your chosen inhibitor looks the most realistic (consider what you learned about substrate recognition and the mechanism of action you observed last week).
   B. Turn off the grid box by clicking on it on the right-side menu. Save an image of the pose you think is best for your inhibitor, both with and without the peptide substrate, using surface representation for the FoCdc14 active site. Use the instructions provided above for saving images and use the file names "Group#_InhibitorDock_alone" and "Group#_InhibitorDock_pep." You will use these in your final report.
   C. Save your PyMOL session at this point with the file name "Group#_Docking."

PART 5. OPTIMIZING TOP INHIBITOR

1. For the last part, you are going to evaluate molecular interactions between your inhibitor and the active site and then design one addition to your inhibitor predicted to enhance binding.
   A. Make sure your view contains just your chosen inhibitor in its preferred pose and FoCdc14 in surface representation. Hide the 01_FoCdc14_Vina prepared structure by clicking and deselecting its line in the right side menu. Make the FoCdc14 surface partially transparent: "Setting" menu → "Transparency" → "Surface" → "20%." Color it uniformly gray: "C" → "grays" → "gray90." Hide the secondary structures if present: "H" → "cartoon." And then show stick structures for the FoCdc14 amino acids: "S" → "sticks." You should see the stick structures now through the transparent surface representation.
   B. Add polar bond labels between the inhibitor and FoCdc14: Wizard menu → "Measurement." The Measurement menu appears in the lower right-side section. In the top drop-down menu select "Polar Contacts." In the main display window you will be prompted to click on the first molecule. Carefully click once on any visible stick of the FoCdc14 structure (avoiding the inhibitor). You should see the entire FoCdc14 structure highlighted with small squares and the prompt changes to "please click on the second molecule . . ." Carefully click on any portion of your inhibitor structure. If there are any polar bonds (H-bonds or ionic contacts) between your inhibitor and the FoCdc14 active site they should be labeled with a yellow dashed line and distance value (in Å). Click "Done" at the bottom of the "Measurement" section. See if you can identify the amino acids in FoCdc14 involved in any polar contacts (if you click on an amino acid you will see its identity in the Command Line window).
   C. Next, color the FoCdc14 surface based on amino acid side chain character. In the Command Line window type: "color green, resn ALA+VAL+LEU+ILE+MET+PRO+PHE+TRP+TYR." This colors the surface green wherever amino acids with mostly nonpolar (hydrophobic) side chains are located. Type "color red, resn GLU+ASP"

and then "color blue, resn LYS+ARG." This colors red and blue, respectively, the surface where amino acids with negative and positive side chains exist.

i. Now, you can see areas where hydrophobic interactions exist or are possible as well as locations for creating electrostatic interactions.

ii. Save an image at this point of your inhibitor bound to the FoCdc14 active site with polar bonds labeled using the title "Group#_PolarBonds."

D. Choose one addition to make to your inhibitor that you predict will increase its binding affinity for FoCdc14. It can be either a nonpolar group in an area where the FoCdc14 surface is green (nonpolar) to create a hydrophobic interaction or a polar group pointing to a red or blue area of opposite charge to create a polar or electrostatic bond.

E. Delete *everything* from the PyMOL session using the right-side menu *except* the Peptide, FoCdc14, and your chosen inhibitor's Vina docking result (Run_#_Vina). Use "A" → "Delete object" on each line. Save your PyMOL session with the name "Group#_DesignStart."

i. Use the Builder tool found at the top right of the menu bar in PyMOL. Click on an atom position in your inhibitor where you want to add a new functional group. Then select the desired group from the Builder window. Your group may take multiple atom additions—that's fine. If you add a charge group, manually set the charge on the appropriate atom using the Builder tool. If you are feeling adventurous, make a couple of changes. You're just exploring now.

ii. Save your new molecule: "File" → "Export Structure" → "Export Molecule." Choose your inhibitor from the Selection drop-down menu (it will be labeled "Run_#_Vina"). Under the PDB options check the first, second, and fourth boxes only and then click "Save . . ." Choose PDB as the file format and use the file name "Group#_OptimizedLigand1."

F. Load your optimized inhibitor back into your PyMOL session using "File → Open." Repeat the docking run in Docking Pie using the exact same procedure describe in Part 4 above except for selecting just your optimized inhibitor as the ligand. You will need to deselect the previous inhibitors in the ligand tab.

G. View the final docking poses for your inhibitor. Save the output table .csv file as before. Compare the affinity value for your optimized inhibitor and your original inhibitor. Is your new molecule predicted to bind FoCdc14 with higher affinity? If you were expecting new bonds to be created, use the Measurement tool again to visualize bonds between FoCdc14 and the inhibitor. Were you successful in creating the intended bond(s)?

i. Save your final PyMOL session with the title "Group#_Optimized."

ii. Save an image of the best pose of your optimized inhibitor bound to FoCdc14 using the name "Group#_Optimized."

H. If time permits, go back and try again if you were not successful the first time. This is a natural part of the drug design process.

DATA ANALYSIS AND SUBMISSION

You need to have the following files stored in your group's data folder for Week 13:

1. The .cif files of your FoCdc14 AlphaFold structure prediction along with the 5XW5 .cif file for the ScCdc14 crystal structure

2. The .pse PyMOL session file and image files titled "Group#_Superpose"
3. The .pse PyMOL session file and image files titled "Group#_FoCdc14_substrate"
4. Five "Inhibitor#.pdb" files generated in Avogadro from the provided SMILES codes
5. "Group#_InhibitorDock_alone" and "Group#_InhibitorDock_pep" image files
6. A single Excel file where you have compiled the docking results tables from the five inhibitors and your final optimized inhibitor
7. The .pse PyMOL session file titled "Group#_Docking"
8. Image file of the FoCdc14 active site with partial transparency showing polar bonds between inhibitor and enzyme title "Group#_PolarBonds"
9. The .pse PyMOL session file titled "Group#_DesignStart"
10. The "Group#_OptimizedLigand1.pdb" file
11. Final .pse PyMOL session file and image file titled "Group#_Optimized"

In addition, submit a single document (Word or PDF) to the course learning management system that contains the following:

1. The RMSD value for the superposition of FoCdc14 and ScCdc14 structures and your conclusion about how similar the structures are
2. Image of the FoCdc14 and ScCdc14 structure superposition and statements about (1) the amino acid regions that do not superpose nicely and (2) whether the active site is among the poorly aligned regions
3. Image of the FoCdc14 protein alone in surface representation showing the core phosphopeptide bound in the active site ("Group#_FoCdc14_substrate" image file)
4. Images of the chosen inhibitor's best pose in the FoCdc14 active site, with and without the peptide substrate
5. Image of the FoCdc14 active site with partial transparency, colored by amino acid side chain property, showing polar bonds between inhibitor and enzyme. Include a list of the amino acids involved in the polar bonds with inhibitor below the image.
6. Images of the optimized inhibitor docked to FoCdc14 with and without the original inhibitor present
7. A comparison of the affinity score for your chosen inhibitor pose and the affinity score for docking of your optimized inhibitor, with a conclusion about how successful you were in increasing the predicted affinity
8. An explanation of why you chose to make that particular modification to your inhibitor and the expected change in binding (i.e., what new bond or interaction did you expect to make?)

NOTES

PREFACE

1. J. Hughes, S. Rees, S. Kalindjian, & K. Philpott, Principles of early drug discovery, *British Journal of Pharmacology 162* (2011), 1239–1249.

INTRODUCTION

1. E. V. Filippova, K. J. Kieser, C.-H. Luan, Z. Wawrzak, O. Kiryukhina, E. J. Rubin, & W. F. Anderson (2016), Crystal structures of the transpeptidase domain of the mycobacterium tuberculosis penicillin-binding protein PonA1 reveal potential mechanisms of antibiotic resistance, *FEBS Journal*, *283*(12), 2206–2218, https://doi.org/10.1111/febs.13738.
2. E. S. Istvan & J. Deisenhofer (2001), Structural mechanism for statin inhibition of HMG-CoA reductase, *Science*, *292*, 1160–1164.
3. N. Amaresan & K. Kumar (2025), *Compendium of phytopathogenic microbes in agro-ecology: Fungi* (Vol. 1, 1st ed.), Springer Cham.
4. D. P. Bebber & S. J. Gurr (2015), Crop-destroying fungal and oomycete pathogens challenge food security, *Fungal Genetics and Biology*, *74*, 62–64, https://doi.org/10.1016/j.fgb.2014.10.012.
5. World Food Programme (2024), *WFP 2025 global outlook*, Rome, https://wfp.tind.io/record/129881?v=pdf.
6. J. Yuen (2021), Pathogens which threaten food security: Phytophthora infestans, the potato late blight pathogen, *Food Security*, *13*(2), 247–253, https://doi.org/10.1007/s12571-021-01141-3.
7. N. Ordonez, M. F. Seidl, C. Waalwijk, A. Drenth, A. Kilian, B. P. H. J. Thomma, R. C. Ploetz, & G. H. J. Kema (2015), Worse comes to worst: Bananas and Panama disease—When plant and pathogen clones meet, *PLOS Pathogens*, *11*(11), e1005197, https://doi.org/10.1371/journal.ppat.1005197.
8. J. Berg, G. Gatto Jr., J. Hines, J. L. Tymoczko, & L. Stryer (2003), *Biochemistry* (10th ed.), Macmillan.

CHAPTER 1

1. J. Jumper, R. Evans, A. Pritzel, T. Green, M. Figurnov, O. Ronneberger, et al., Highly accurate protein structure prediction with AlphaFold, *Nature*, 596 (2021), 583–589, https://doi.org/10.1038/s41586-021-03819-2.

CHAPTER 3

1. J. L. Hartley, G. F. Temple, & M. A. Brasch (2000), DNA cloning using in vitro site-specific recombination. *Genome Research*, *10*(11), 1788–1795.
2. D. G. Gibson, L. Young, R. Y. Chuang, J. C. Venter, C. A. Hutchison III, & H. O. Smith (2009), Enzymatic assembly of DNA molecules up to several hundred kilobases, *Nature Methods*, *6*(5), 343–345.
3. B. Zhu, G. Cai, E. O. Hall, & G. J. Freeman (2007), In-Fusion™ assembly: Seamless engineering of multidomain fusion proteins, modular vectors, and mutations, *Biotechniques*, *43*(3), 354–359.

4. F. Studier (1998), Use of T7 RNA polymerase to direct expression of cloned genes, *Methods in Enzymology, 185*, 305–313.
5. L. M. Guzman, D. Belin, M. J. Carson, & J. Beckwith (1995), Tight regulation, modulation, and high-level expression by vectors containing the arabinose PBAD promoter, *Journal of Bacteriology, 177*(14), 4121–4130.

CHAPTER 4

1. M. M. Bradford (1976), A rapid and sensitive method for the quantitation of microgram quantities of protein utilizing the principle of protein-dye binding, *Analytical Biochemistry*, *72*(1–2), 248–254.

CHAPTER 8

1. L. M. Siegel & K. J. Monty (1966, February 7), Determination of molecular weights and frictional ratios of proteins in impure systems by use of gel filtration and density gradient centrifugation: Application to crude preparations of sulfite and hydroxylamine reductases, *Biochimica et Biophysica Acta*, *112*(2), 346–362.

CHAPTER 10

1. L. Michaelis & M. Menten (1913), Die kinetik der invertinwirkung, *Biochemische Zeitschrift, 49*, 333–369.
2. W. Q. Wang, J. Bembenek, K. R. Gee, H. Yu, H. Charbonneau, & Z. Y. Zhang (2004, July 16), Kinetic and mechanistic studies of a cell cycle protein phosphatase Cdc14, *Journal of Biological Chemistry, 279*(29), 30459–30468.

   A. G. DeMarco, K. L. Milholland, A. L. Pendleton, J. J. Whitney, P. Zhu, D. T. Wesenberg, et al. (2020), Conservation of Cdc14 phosphatase specificity in plant fungal pathogens: Implications for antifungal development, *Scientific Reports, 10*(1), Article 12073.

CHAPTER 11

1. S. C. Bremmer, H. Hall, J. S. Martinez, C. L. Eissler, T. H. Hinrichsen, S. Rossie, et al. (2012, January 13), Cdc14 phosphatases preferentially dephosphorylate a subset of cyclin-dependent kinase (Cdk) sites containing phosphoserine, *Journal of Biological Chemistry, 287*(3), 1662–1669.

   C. L. Eissler, G. Mazon, B. L. Powers, S. N. Savinov, L. S. Symington, and M. C. Hall, (2014, April 10), The Cdk/Cdc14 module controls activation of the Yen1 Holliday junction resolvase to promote genome stability, *Molecular Cell, 54*(1), 80–93.

   A. G. DeMarco, K. L. Milholland, A. L. Pendleton, J. J. Whitney, P. Zhu, D. T. Wesenberg, et al. (2020, July 21), Conservation of Cdc14 phosphatase specificity in plant fungal pathogens: Implications for antifungal development, *Scientific Reports, 10*(1), Article 12073.

   C. H. Gray, V. M. Good, N. K. Tonks, & D. Barford (2003, July 15), The structure of the cell cycle protein Cdc14 reveals a proline-directed protein phosphatase, *EMBO Journal, 22*(14), 3524–3535.
2. C. L. Eissler, G. Mazon, B. L. Powers, S. N. Savinov, L. S. Symington, and M. C. Hall, (2014, April 10), The Cdk/Cdc14 module controls activation of the Yen1 holliday junction resolvase to promote genome stability, *Molecular Cell, 54*(1), 80–93.

CHAPTER 13

1. J. Jumper, R. Evans, A. Pritzel, T. Green, M. Figurnov, O. Ronneberger, et al., Highly accurate protein structure prediction with AlphaFold, *Nature,* 596 (2021), 583–589, https://doi.org/10.1038/s41586-021-03819-2.
2. Schrödinger LLC, *The PyMOL Molecular Graphics System*, Version 3.1.6.1.

ABOUT THE AUTHOR

MARK C. HALL has served as a professor in the Department of Biochemistry at Purdue University since 2004. In addition to directing a research program on fungal diseases, he developed and has taught an analytical biochemistry course for more than a decade. He also created a related curriculum for the Summer Science Program, a nonprofit organization providing high school summer research experiences. These efforts contributed to Hall's election to Purdue University's Book of Great Teachers in 2023 and receipt of the College of Agriculture's PK–12 Emerging Faculty Impact Award in 2021.

www.ingramcontent.com/pod-product-compliance
Ingram Content Group UK Ltd.
Pitfield, Milton Keynes, MK11 3LW, UK
UKHW060311060926
475094UK00007BA/215

9 781626 713864